I0477174

This Polluted World

We Live In

Gets Worse

By

Andrew D. Anderson

This book is a work of non-fiction. Names and places have been changed to protect the privacy of all individuals. The events and situations are true.

© 2002, 2003, 2004 by Andrew D. Anderson. All rights reserved.

No part of this book may be reproduced, stored in a retrieval system, or transmitted by any means, electronic, mechanical, photocopying, recording, or otherwise, without written permission from the author.

First published by AuthorHouse 04/27/04

ISBN: 1-4107-5784-6 (e-book)
ISBN: 1-4184-5417-6 (Paperback)
ISBN: 1-4184-5418-4 (Hardcover)

This book is printed on acid free paper.

Acknowledgment

I am most grateful to my dear wife Bessie and her help, support, cooperation and patience, my son Nathan and wife Joyce that planted the seed and provided encouragement and research material, and my grandchildren, Danielle and Eric in Merrimack, New Hampshire.

Andrew Anderson

iv

Preface

We do not see or understand our challenge of living in harmony with the earth and its assets. That relationship changed fundamentally the 20th century with the combustion engine when society found itself able to exploit on a massive scale the energy locked in such fossil fuels, coal, oil, electricity, and gas. Humankind has found itself in uncharted territory with respect to energy use, population growth, and lost of fresh air which has been an existing commodity free to all.

This book is a learning tool, and a reference of the broad range of major things that are polluting our planet, depleting our water supplies, originating new diseases that travel in the atmosphere, and the nuclear waste that is temporarily stored and at one site leaked over one million gallons contaminating our water. The United States spent over $3 billion that did not stop the leaks. It is intended as both a refreshing and learning tool to assist anyone in the whole world to realize some of the things that are robbing us of a quality life with a questionable future.

The atmosphere is filled with toxins, the oceans fished to exhaustion, our climate and weather appears to be angry with us and our food and water is being secretly limited. Imagine a future of relentless storms and floods, islands and heavily inhabited coastal regions inundated by rising sea levels, fertile soils rendered barren by drought and the deserts advancing, mass migrations of environmental refugees and armed conflicts over water and other precious natural resources.

We need to realize the world has problems that affect the unborn child and keep on affecting age groups until the elderly is reached. Not only are they affecting us where we live, but the prevailing winds carry microbes and pathogens to different parts of the world.

New research shows that harmful effects of dirty air extend even into the womb of a pregnant woman. Over a dozen studies in the United States, Brazil, Europe, Mexico, South Korea, and Taiwan link dirty air to low birth weight, premature births, still births, and infant deaths.

What is happening to the water? In North America the Colorado River barely makes it to the Gulf of California, and last year even the Rio Grande dried up before it merged with the Gulf of Mexico. In Central Asia the Arial Sea shrunk by half after the Soviet's began diverting water for cotton and other crops. China's Yellow River barely trickles in its lower reaches-and in recent years has gone dry due largely to heavy irrigation upstream. The once mighty Nile, Ganges, barely reaches the sea in dry seasons. Lake Chad in Africa in 1962 was the fourth largest water body on the continent. Now it has shrunk to one 20th of that size. California has seen more than its share of water and will soon have to decide whether to water its farms or its burgeoning cities.

We forget how basically important water is to life. Our bodies are 65 percent water, our blood is 90 percent water, and this planet has over 70 percent of water, yet we treat it as being unlimited. Each year 40 million tons of raw sewage, chemical wastes, fertilizer, herbicides, pesticides, animal feces associated with large scale factory farming, landfills, hazardous waste and other poisons pollute our drinking water, ground water, surface water for our lakes and rivers, and irrigation that waters our growing plants. By 1980, twenty two years ago, the United States took over 24 trillion gallons of water from the ground. To get a better idea of that size number, that's 24,000,000,000,000 gallons of water. Light travels about 6 trillion miles a year in a vacuum or in four years it should travel about 24 trillion miles. Considering the speed of light traveling for four years, that should give the reader an idea of that size number equated to gallons of water.

Wetlands are precious ecological resources. They nurture wildlife, purify polluted waters, check the destructive power of floods and storms, and provide all kinds of recreational activities. They also reduce flood crests and flow rates after rainstorms. During two decades covered by the study, 6 million acres of forest wetlands, 400,000 acres of shrub swamps, 4.7 million acres of inland marshes, and 400,000 acres of coastal marshes and mangrove swamps were destroyed. More than 11 million acres has disappeared.

The offending microbe can survive the traditional heating and cooling techniques we once thought did away with them, but food borne pathogens, bacteria, viruses, and parasites with a potential to harm or kill us are the great hazards today. A single bacterium under the right conditions, divides rapidly enough to produce colonies of billions over the course of a day, and even lightly contaminated food can become highly infectious. It is widely known that they can hide and multiply on sponges, dish towels, cutting boards, sinks, knives, and countertops where they can be easily transferred to food or hands. They can be easily killed by letting them boil in the microwave ovens.

Another form of pollution we cannot stop drifting with the suspended dust particles are soil pollutants such as herbicides, pesticides, a number of microorganisms, bacteria, viruses, fungi, animal feces and billions of metric tons of sediment-borne bacteria moving around the planet. Dust storms above northwestern Africa or the Sahara Desert that covers the Republic of Mali can carry the sediments to the United States and Canada.

Anything in the dust particles are so small that once they are inhaled into the lungs they cannot be exhaled. Some of the contaminants are endocrine disrupters (pesticides and polyaromacic hydrocarbons), carcinogens (dioxin and radioactive isotopes) and some are toxic to cells (heavy metals).

Factory farming has a new way of raising animals. Broiler chickens are bred to gain weight rapidly in sheds containing 10 to 14 thousand chickens each. The big change in raising cattle is in feed lots where they are jammed together and fattened the last six months before slaughtering, standing in deep black feces and when they reach the slaughter house, they are covered with feces and crowded together.

Farmers have been adding antibiotics to animal feed, and some scientist believes the volume of antibiotics used in animal feed equals or exceeds that used in human medicines which are serious threats to human health. They believe it creates strains that are resistant to antibiotics used in human medicine.

One thousand hogs live in separate sheds in varying parts of America and produce 280 billion pounds of manure annually. Ground water washes waste from the hogs, cattle, and chickens polluting rivers, lakes, drinking water, and killing fish.

Bad management of agriculture leads to soil stalinization and degradation and a projection of a shortage of food. Biotech plants or genetic modification may help.

Energy demand nearly doubled in the past three decades and is expected to increase another 60 percent by 2020. Cleaner renewable energy is growing fast supplying nearly 10 percent of the world's total energy from alternative sources as wind turbines, solar cells, biomes fuels, and hydrogen fuel cells that should provide half of the world's energy needs by 2050.

India has a boom in wind power for electricity because the government has made it easier for entrepreneurs to get their hands on the necessary technology and Japan is giving financial incentives for consumers who buy environmentally sound cars. A Shirley, NY company has filed an application to erect a cluster of 10 wind turbines a mile offshore along the east New England coast producing enough electricity for 600 homes. Oregon, Texas, Arizona, and California are other states that are using wind power.

California Governor Gray Davis signed a law requiring automakers to cut their cars' carbon emissions by 2009 and this government is giving tax incentives for cutting back on energy use.

Because this book covers a broad range of topics relevant to the general subject of pollutants we now have, each reader at varying times may have specific interest. The text is divided into chapters and parts with page numbers to help locate that information. Each gives direct clues to what is in the chapter to help the reader and the prospective buyer that scans the Contents to reinforce what it covers.

Andrew D. Anderson

Contents

Chapter 1: Pollution in America

Introduction

The purpose of this book is not to stimulate, generate, or introduce fears. It is to make people aware of existing and growing pollutants that contribute and restrict the freedom of our lives as we continue to focus on our environments. When we look or concentrate on things that control our lives we react in such a way that will help prolong our lives.

Greenhouse Effect

For millions of people around the world, simply breathing has become risky business. An estimated three million people die annually from the effects of pollutants, sulfur dioxide, particulate, nitrogen, and ozone, among others that result from the burning of fossil fuels. Children in the developing countries are the hardest hit, often inhaling two to eight times the amount of pollutants deemed safe by the World Health Organization.

Worldwide warming is included because it is caused by burning of forest fires and pollution from vehicles and airplanes that keeps growing with no end. It is increasing and is caused by humans as more toxins enter the atmosphere. Research indicates one effect of global warming is caused by burning fossil fuels. There are pollutants that harm the health of many people, but they do not realize it. If all toxicants stopped being released in the atmosphere, it would remain there for decades. With the populations and the number of vehicles growing, the most America can do is slow the increasing ozone levels. Riding trains will not solve the worldwide warming, but no single action will. Riding trains will help in the heavily populated urban areas where more people will have lung damage.

1

Andrew D. Anderson

What is this global warming and greenhouse effect? The greenhouse warming effects are directly from the sun that is very hot and earth that is relatively cool. It starts when our sun floods earth with a bath of warm energy. The atmosphere, acting like an insulating blanket or the glass on top of a greenhouse, keeps some of that heat from escaping back into space. The results are a temperature regulation system that keeps things just right for life, not too hot and not too cold. Human activity has increased the proportion of certain atmosphere components called greenhouse gases and the result is a rise in global temperatures held near the earth by the atmosphere.

One gas that has been a chief malefactor in the greenhouse story is carbon dioxide. The rise in carbon dioxide is due in part to the increased combustion of fossil fuels. It is the single largest waste product of modern society. That added to more than five billion tons of carbon from fossil fuels in 1988.

Green plants are a principal consumer or a sink for carbon dioxide. That is why burning tropical forest for agriculture, cattle, and cooking fuel contributes. The Brazilian fires, fires in the Northeast and in California, Oregon, Arizona, Canada, and Colorado in the United States put significant amounts of carbon dioxide into the air.

Burning tropical forests to provide room for more cattle ranches, changes from one to a triple effort. Large fell forest trees make a wonderful home for termites, and the end results are the same as cows and methane.

Methane is the second most important greenhouse gas we are adding to the atmosphere after carbon dioxide. Three of the sources are emissions from solid waste dumps, coal mining, and natural gas leaks. Large amounts of methane originate in a surprising place, the digestive tract of cows. Between 3 and 10 percent of cattle fodder, cattle food, is turned into methane. It has been estimated that there are 1.2 billion heads of cattle on earth and all of them belch furiously. The average cow belches up to 400 liters of methane a day, and the yearly global contribution is about 1 million tons from cows alone.

2

Cattle farms produce 1.4 billion tons of animal waste which contaminate drinking water that kill fish and spread disease. Cattle dung also emits methane.

Many people will be surprised by the relatively high level of health risk associated with air toxins in their area said Bill Pease, a toxicologist at the New York based Environmental Defense Fund who obtained the Environmental Protection Agency results. There are a lot of things we do every day, but do not realize it has a toxic consequence as dry clothes cleaners and businesses. There are 148 chemicals such as formaldehyde, carbon, tetrachloride, benzene, chloroform, ethylene, dichloride, and methyl chloride in the air.

Most of these chemicals are not recognized by us, but Americans increase their risk of cancer when they draw breath outdoors. A place to escape toxic chemicals in the air "does not exist." The Environmental Protection Agency found that every American inhales unsafe levels of at least eight chemicals and 20 million Americans face at least a 10,000 life time risk of getting cancer a result of a 1990 Cumulative Exposure Project.

Airplanes are another producer of pollution at busy airports all over this world. We usually think of pollution coming from vehicles, but Jets emit tons of oxides every year that is an ingredient in acid rain that harms trees and wildlife. Logan Airport ranks sixth on the list of the worst emitters of smog and Robert Durand predicts Logan airport will rank number one by 2010. The newest Jets are quieter, but their redesigned engines release more nitrogen oxide. Logan Airport and all of the bigger airports also cause vibration and noise.

Airplanes spew out nearly 4 million tons of nitrogen oxide, much of it while cruising in the troposphere five to seven miles above the earth where ozone is formed. Ozone experts estimates that air traffic accounts for 8 percent of all global greenhouse warming. They are concerned that jet emissions of carbon dioxide and nitrogen oxides could be a concern in the near future. The Office of Technology Assessment estimated that aircraft emissions represent about five percent of the 1.4 million tons of air pollution produced annually from all sources. Changing to electric engines for cars, and carts that service airplane fleets, buses, food trucks, and baggage-handling carts will reduce emissions.

Air pollution makes people sick and damages the environment and trees, lakes, and all life. Air pollutants have made the ozone layer around the Earth thinner, leading to increase skin cancer because of more ultra violet rays. It also dirties buildings and statues, and some pollutants eat away the stone in buildings, monuments, and statues. Air pollution also causes haze that can make it harder to see and it comes from burning fossil fuels and the use of chemicals alter the earth's chemistry and threaten the food, water, and air supplies we depend on.

Air pollution has three principal man-made sources, energy use, vehicular emissions, and industrial production which will continue to increase with economic and population growth. In 1998, approximately 113 million Americans lived in areas that failed to meet pollution standards. Throughout the world, poor air quality contributes to hundreds of thousands of deaths and diseases each year. In Paris, France, about 3 million cars enter the capital daily, shrouding the Eiffel Tower in smog and crowding emergency rooms with people suffering from bronchial distress. Air pollution including acid rain, devastates forests, crops, and waterways, and works its way into the water cycle and food chains.

It is related to a number of respiratory diseases, including pulmonary, bronchitis, emphysema, bronchial asthma, lung cancer, eye irritation, weakened immune system and premature lung tissue aging. The American Lung Association estimates the annual health costs of exposure to the most serious air pollutants is $40 to $50 billion. Also lead contamination causes neurological and kidney disease that can be responsible for impaired fetal and mental development.

There was an "Asian Brown Cloud" two miles-thick blanket of pollution over South Asia in 2002 and may be causing the premature deaths of a half-million people. In India deadly flooding and drought in different areas occurs each year. Paul Crutzen, a scientist at the Max-Planack Institute for Chemistry in Mainz, Germany says, when we think about air pollution many think of industry and fossil fuels as the only causes, but often ignore burning, including forest fires and the burning of vegetation to clear land or to warm the homes of poor people.

The heavy cloud of pollution is also caused by auto emissions, factories, and waste incineration which reduce the amount of sunlight reaching the ground and the oceans by 15 percent to 20 percent, cooling the land and water while heating the atmosphere. The study was found by using data from ships, planes, and satellites.

Clean air, if it could be found, would have to be void of human habitation and not affected by violent natural phenomena. It would contain water vapor, suspended solid particles, and many rare gases, but they would all be integrated in rightful amounts created by the creator.

Smog

Smog is a word made up from smoke and fog, and probably the most well-known form of air pollution. However, it can be good or bad, depending on where it is located. When ozone is high in the atmosphere, it shields the ultra-violet light coming from space and protects human health and the environment. When the ozone is at ground level, it becomes the most harmful element in smog.

Ground-level ozone is produced by the mixing of pollutants from many sources, including smokestacks, vehicles, lawn mowers, construction equipment, and paints. When a car burns gasoline or a painter paints a house, dangerous fumes rise into the air.

Often wind blows smog-forming pollutants away from their sources. The reaction that creates smog occurs while the pollutants are being blown through the air by the wind. This is why smog is often more serious miles away from where the pollutants were created. The smog forming pollutants are brought together in the sky, and if it is hot and sunny, smog forms more easily.

Ozone is formed in the stratosphere when ultraviolet radiation splits diatomic molecules of oxygen into two atoms. The concentration of ozone at different altitudes can affect the movement of ultraviolet rays through the atmosphere, which in turn influences the radioactive and meteorological processes that determine weather conditions. Thus, if the ozone equilibrium is disturbed, major environmental changes occur.

Photochemical air pollution, also commonly known as smog, has become important since the middle 1940s primarily because of the increase in vehicle emissions. This type of pollution is the result of a number of complex chemical reactions.

The development of ozone as a photochemical oxidant occurs primarily in urban areas. In urban areas it is a primary concern. It has a daily cycle which increases during the daylight hours and rapidly decreases during the night periods. In contrast, the ozone in rural areas might persist at higher altitudes with little variation over a 24 hour period. The night time ozone is probably associated with nocturnal inversions produced near the ground and these inversions reduce the mixing with nitrogen oxide and hydrocarbons and reduce the contact with the surface. Above the inversion layer the half-life of ozone may be as long as 80 hours.

Weather and location determine where the smog goes and how bad it will become. When temperature inversions occur, the warm air stays near the ground instead of rising, and the winds are calm. Smog may stay near the ground for days at a time as traffic and other pollution sources add more pollutants to the air and the smog gets worse.

People connect dirty air with cities and areas around them, but in major industrial nations such as the United States, smog is not just limited to the cities. On a clear day, the Grand Canyon is one of the most beautiful sites in the world. On some days it is difficult to see the bottom of the Canyon. On the other side of the continent, the Great Smoky Mountains, another beautiful place in the United States, also suffers from smog.

Damages the air does to buildings and statues indicate how strong it is. Most television and radio weather reporters in major cities include ozone and air quality reading as part of the daily weather statistics.

It has altered the region's monsoon rains, increased rainfall and flooding in Bangladesh, Nepal, and Northeastern India, and reduced the needed seasonal precipitation in Pakistan and India. The pollution may be cutting India's winter rice harvest by 10 percent and causing drought, sunlight reduction, floods, and acid rain that can hurt agricultural yields. Prevailing winds push pollution clouds halfway round the world in a week's time moving microscopic suspended particles of pollutants called aerosols.

Contribution to Death

Throughout the world, poor air quality contributes to hundreds of thousands of deaths and diseases each year, dying forests and lakes, and corroding buildings and monuments.

The nation's worst smog problems are located in Southern California, Chicago, Houston, Texas, and New York metropolitan areas. Eleven Atlantic Coast states, Maine, New Hampshire, Vermont, New York, Pennsylvania, Massachusetts, Rhode Island, Connecticut, New Jersey, Maryland, Delaware, and the District of Columbia came under regulatory order to control smog due to their high levels of ozone. Mexico City has the worst air in the country. It has become so serious that schools have closed for days, factory and vehicle use have been halted, and thousands of people have become ill. A unique solution to dissipate the smog is six 20-foot fans to be situated around the city to break the entrapment of air and allow the smog to rise.

Long-term exposure to air pollution significantly raises the risk of dying from lung cancer. The researchers found that the number of lung cancer deaths increased 8 percent for every increase of 10 micrograms. Co-author George Thrust, an NIGH Environmental Scientists said, the biggest sources of such pollution are coal-burning power plants in the Midwest and East and diesel trucks and buses in the West. He also said the lung cancer risks were virtually identical to those faced by nonsmokers who live with smokers and are exposed long-term to second hand cigarette smoke according to a study of a half-million people in some of the biggest United States metropolitan areas. The risk is what scientists call combustion-related fine particulate matter, or soot emitted by vehicles, coal-fired power plants, and factories.

According to the American Lung Association, Knoxville ranks eighth among the most ozone polluted cities. It is worse than New York City. Great Smoky Mountains National Park, one of the most popular national parks in the nation, is also one of the most polluted, because of both ozone and the fine particles in the air that dramatically reduce visibility.

Both of the types of air pollution, ozone and fine particles increasingly worry doctors. Ten or 20 years ago, people believed there was a safe level for air pollution Dr. George Thurston said as an associate professor at the New York University School of Medicine. Currently levels of particle pollution are increasing the risk of death from lung cancer who would not otherwise die he says.

This fine-particle pollution comes from sulfur and some nitrogen oxides emitted by older electrical generating plants, diesel engines, and other sources. It also causes summertime "haze" as the particles absorb and reduce light. These toxic particles can lodge deep in the lungs, but until recently it has not been clear just how breathing bad air causes cardiovascular problems.

The heart and the lungs are actually a single unit, says Douglas Dockery of the Harvard School of Public Health. Blood goes to the right side of the heart, which pumps it to the lungs, and it goes out the left side of the heart. Recent research has shown that fine particles interfere with what Dockery calls "the electrical control of the heart," the heart's ability to vary and adjust its rate in response to exercise or stress.

Robert Brook at the University of Michigan, demonstrated that fine particles mixed with ozone, comparable to dirty urban air, cause blood vessels in the arms of healthy adults to constrict which is a good marker of what happens in the heart he says. Chemical processes similar to those that form fine particles also create ozone when heat and sunlight act on emissions from vehicles, power plants, and factories. Despite a 29 percent drop in levels of all pollutants combined since the Clean Air Act, ozone has actually worsened in the northern and southeastern United States over the past 10 years.

There are a huge number of new cases of asthma. An estimated 24.7 million Americans have been diagnosed with asthma, according to the American Lung Association and the disease costs the nation $12.7 billion yearly. In children under age 4, asthma incidence jumped by 160 percent. Rob McConnell, a University of Southern California researcher said, nobody really understands why.

They looked at children exposed to ozone in 12 California towns, and found that kids who play three or more team sports in communities with high levels of ozone are three times as likely to develop asthma as those who do not play sports breathing less outside air. His group also demonstrated that small increases in ozone, about one sixth of the current standard, resulted in a 63 percent jump in school absences for respiratory illnesses. Among children who already have asthma, there is roughly a doubling of symptoms when they live in communities with dirty air, McConnell says. Almost half of those kids are sick much of the time.

Lung Cancer

The study, published in today's Journal of the American Medical Association, involved 500,000 adults who in 1982 enrolled in an American Cancer Society Survey on cancer prevention. The researchers examined participant's health records through 1998 and analyzed data on annual air pollution averages in the more than 100 cities in which participants lived. Lung cancer death rates were compared with average pollution levels, as measured in micrograms per cubic meter of air.

Allen Derry, a scientist at the National Institute of Environmental Health Sciences, which funded the study called it, "the best epidemiological evidence that we have for that type of exposure associated with lung cancer death." Co-author George Thrust, an NIGH environmental scientist said, this study is compelling, because it involved hundreds of thousands of people in many cities across the United States who was followed for almost two decades. Thurston said the lung cancer risks were virtually identical to those faced by nonsmokers who live with smokers and are exposed long-term to second hand cigarette smoke. Thurston said that annual averages for fine-particulate pollutants have fallen significantly since the early 1980s. But in the 1999-2000 periods, averages were still at or above the EPA limit in metropolitan areas such as New York, Washington, Chicago, and Los Angeles.

The environmental Protection Agency set average annual limits at 15 micrograms per cubic meter in 1997, and then it tightened its standards to include fine particulate matter, or pollutants measuring less than 2.5 micrometers. That is about one-28th the width of a human hair. The regulation adopted after a study linked fine particulate pollution with lung cancer.

Carbon Monoxide and Lead Poison

Carbon monoxide comes from the burning of gasoline, natural gas, wood, coal, and oil. It is a dangerous gas that reduces the ability of blood to carry oxygen to the body's cells. In 1940, cars and trucks created about 28 percent of the carbon monoxide emissions, while homes burning coal and oil created about 50 percent. From 1940 through 1970 emissions from cars and trucks nearly tripled and by

1970, cars and trucks accounted for 71 percent of all carbon monoxide, and a dozen years later in 1982, they produced 80 percent of the total carbon monoxide emissions. We cannot do without cars or trucks, but we can reduce the number that keeps growing.

Lead poisoning is the most common and most devastating environmental disease affecting young children, according to the Centers of Disease Control. Lead can damage the brain and nervous system. Until recently, the main source of lead pollution was leaded gasoline, and lead-based paint was widespread in housing in the United States. HUD estimated that 57 million, or about three-fourths, of the 77 million privately owned and occupied homes built before 1980 contained lead-based paint. Almost 10 million of these homes were occupied by families with children under 7 years of age, who are the most vulnerable to lead poisoning.

Nitrogen Dioxide and Sulfur Dioxide

Nitrogen dioxide also comes from the burning of fuels such as gasoline, natural gas, coal, and oil. Cars and trucks are also a major source of nitrogen dioxide. Nitrogen dioxide is a major part of smog and causes the same health and environmental effects as smog. Nitrogen dioxide is also in acid rain, which damages trees, lakes, and eats away buildings and statues.

Particulate matter is the dust, smoke, and soot that come from the burning of fuels from industrial plants, from farmlands, and from unpaved roads. Particulate matter can irritate the nose, throat, and lungs and when particulate matter hangs in the air, it creates a haze. From 1940 to 1971, particulate matter generally increased. Pollution control laws led to a drop in the matter most of which occurred during the 1970s. Its emissions dropped from almost 61,000 short tons in 1988 to 43,000 short tons in 1993.

Sulfur dioxide comes from the burning of coal and oil. It can lead to serious breathing problems and is also a major part of acid rain which can also damage trees and lakes. From 1940s to the 1970s, sulfur dioxide emissions increased as a result of the growing use of fossil fuels, especially coal by industry and electrical utility plants. It has dropped because of cleaner fuels with lower sulfur content and the greater use of pollution devices such as scrubbers that clean the factory emissions. Most emissions today come from the burning of fuels.

Oil, auto industries, and legislators have concentrated on miles per gallon fuel efficiency. They have shown that they can adapt to tighter emission standards when they have to. Lighter weight engines have been introduced, fuel-injection, catalytic converters, and other improvements. It would be much better if they eliminated emissions and in promoting alternative transportation, such as mass transit systems and commuter rail.

The government permitted the industries to make a lighter weight car to aid in mpg, but the body structure of the vehicles cannot provide much safety upon impact. Thus for better vehicle mileage we decreased the length of our lives when ever there is an auto or truck accident. It really does not matter what type of accident that occurs on the highways, the vehicle is usually completely destroyed and lives lost. What a price to pay for better gasoline mileage.

The standards required each domestic automaker to increase the average mileage of the new cars sold every year until achieving 27.5 miles per gallon by 1985. Under the rules, car manufacturers could still sell the big, less efficient cars with powerful 8 cylinder engines, but to meet the "average" fuel efficiency rates, they also have to sell smaller, more efficient cars. The automakers that failed to meet each years standards are required to pay fines. Those that surpassed the rates earned credits that they can use in years when they fall below the requirements.

Use of Gasohol, Hydrogen, and Electricity

To help reduce the nation's dependence on imported oil in 1991, Congress enacted the National Defense Authorization Act (PL

101-510) which included a provision directing federal agencies to purchased gasohol, gasoline containing 10 percent ethanol when it was available at prices equal to or lower than gasoline. Executive Order 12759 of April 1991 required federal agencies operating more than 300 vehicles to reduce their gas consumption by 10 percent as an incentive to use gasohol. Gasohol cost more than gasoline because it includes ethanol and gasohol that is sometimes unavailable due to the high cost of transporting and storing it.

Factories that convert corn into the gasoline additive ethanol are releasing carbon monoxide, methanol, and some carcinogens at levels many times greater than they promised the government. In April 2002 a letter to the Environmental Protection Agency said the problem is common to most, if not all, ethanol facilities.

The ethanol industry is pressing to significantly expand production, as many states phase out another widely used fuel additive, MTBE because it is polluting water supplies. Volatile organic compounds being released by the ethanol plants include formaldehyde and acetic acid, both carcinogens. Methanol, although not known to cause cancer, also is classified as a hazardous pollutant.

Recent test have found emissions of volatile organic compounds ranging from 120 tons a year, for some of the smallest plants and up to 1,000 tons annually for some of the bigger plants. States started measuring emissions of volatile organic compounds at ethanol plants about a year ago, after complaints of foul odors. One facility in St. Paul had to install pollution control equipment worth $1 million to reduce those emissions.

For decades, advocates of hydrogen have promoted it as the fuel of the future, abundant, clean, and cheap. Hydrogen researches from universities, laboratories, and private companies claim their industry has already produced vehicles that could be ready to market if problems of fuel supply and distribution could be solved. In 1995, five of 11 passenger carts in use at the Palm Springs CA airport were powered by hydrogen. Other experts contend that economics and safety concerns limit hydrogen's wider use.

Electric Cars must be recharged to often, or they are too expensive. The primary difficulty with electric cars lies in inadequate battery power. Currently, these cars use lead-acid or nickel-cadmium batteries and have a range of 50 to 100 miles on a single charge. This range is reduced by factors such as cold temperatures, the use of air condition, vehicle load, and steep terrain. Recharging the battery generally takes about eight hours, although recent technological development will likely lower this to 4 to 6 hours. The Big Three automakers, along with the United States Department of Energy, formed the United States Advanced Battery Consortium in 1991. It was a $260 million effort aimed at researching several potential batteries for the electric car. In addition, researchers suggested that emissions from mining, smelting, and recycling the lead needed for a large fleet of electric vehicles would pose serious threats to public health.

Tax breaks became available in July 1993 for people who bought cars that were powered by alternative energy sources, especially for electric cars. The law provided for a credit of 10 percent of the price, up to $4,000 for electric vehicles. The law also permitted a deduction of up to $2,000 for other clean-fuel vehicles. The tax breaks were intended to compensate for the price difference and to jump-start production of the vehicles. In 1998, 10 percent of all new cars offered for sale in California were to be nonpolluting.

Other Polluting Vehicles

In 1994, the EPA reported that an estimated 10 percent of the nation's air pollution is generated by lawn and garden equipment, lawn mowers, chain saws, and golf carts. Carol Browner of EPA estimated that Americans use 89 million pieces of such equipment, with lawn mowers alone accounting for 5 percent of the nations pollution. The agency established engine labels and warranty requirements, exhaust emissions standards, and test procedures, requiring that engine makers met the new requirements by 1996. Agency officials predicted the new regulations would reduce smog forming hydrocarbon emissions by 32 percent and carbon monoxide by 14 percent by year 2003.

The most widespread technological invention to reduce emissions has been the introduction of electrostatic precipitators and filters to control emissions from power plants. They will reduce particulate emissions from smokestacks by 99.5 percent, but they do nothing about gaseous emissions. The primary techniques used to reduce sulfur dioxide has been the use of scrubbers that remove 95 percent of the residue. In general, new power plants are tightly regulated in most industrialized countries. The problem comes in modifying existing facilities. However, the available technologies have their own inadequacies. They create scrubber ash, hazardous waste, but do nothing to control carbon dioxide emissions on existing facilities.

Health expert Thomas Crocker of the University of Wyoming estimated the cost to be as high as $40 billion annually in health care and lost productivity. The American Lung Association calculated the direct and indirect health care cost associated with air pollution to be $16 billion to $40 billion a year.

Pollution Harmful to Babies

Gary Polakovic, Los Angeles Times Studies, link smog with infant deaths. Scientists have long known that the extreme levels of air pollution found in the developing world can harm babies, and that lesser pollution in United States cities can sicken or kill the elderly and infirm.

The new research shows that the harmful effects of dirty air can extend even into the womb. More than a dozen studies in the United States, Brazil, Europe, Mexico, South Korea, and Taiwan have linked smog to low birth weight, premature births, still births, and infant deaths.

In the United States, the research has documented ill effects on infants even in cities with modern pollution controls. The findings have helped prompt California officials to seek more stringent smog control. A study conducted by Beate Ritz, an epidemiologist at the UCLA Center for Occupational and Environmental Health, links for the first time air pollution and birth defects in Southern California.

We know there are serious health effects from low levels of air pollution, said Aaron Cohen, an epidemiologist and principal scientist for the Health Effects Institute in Boston. UCLA found that women exposed to high levels of ozone and carbon monoxide were three times more likely to have babies with cleft lips and palates, and defective heart valves.

Three researchers looked at thousands of pregnant women in the Los Angeles area from 1987 to 1993 and compared those living in areas with relatively dirty air to those living in cleaner areas. Virtually the entire study area met federal standards for carbon monoxide, and most of the regions complied with ozone requirements.

The study found that greatest risk occurred during the second month of pregnancy, when a fetus gains most of its organs and much of its facial structure. Scientists from Harvard School of Public Health and the University of Basel in Switzerland concluded that as much as 11 percent of infant deaths in the United States might be a result of microscope particles in the air.

The Clean Air Act regulates smog levels to protect certain sensitive groups, including children, the elderly, and people with respiratory ailments. Pollutants inhaled by pregnant mothers can reach fetuses through the umbilical cord, research has found.

Electronic Smog

There is a new kind of smog that is adding to the destruction of mankind. Scientists are beginning to take a hard look at a form of pollution that is pervasive throughout the country, although it cannot be seen, heard, felt, or smelled. This newest pollutant is "electronic smog." It is caused by the increase of low level radiation that emanates from radio and TV broadcasting equipment, microwave ovens, citizen-band radios, power lines, computers, and radar.

Many scientists believe that electronic pollution is at a low enough level not to be harmful to humans. Others are concerned that as it worsens it will bring many health problems. It could cause ailments ranging from behavioral disorders to stunted growth, cancer, and chromosome damages. The danger has advanced since WW II when there were six TV stations. Today, 1,000 stations feed 120 million TV sets and most households have microwave ovens.

Every time this author walked under a power line at a mall in Massachusetts I immediately developed a severe headache. When I walked away a few feet, it disappeared.

Electromagnetic interference is a major problem and will only get worst creating technical, legal, health, and social problems. It will only grow worse as electronic devices become more widespread, said Bureau Director Earnest Ambler.

Computer Smog in China

When low-grade plastics and computer screens were burned and dumped beside rice paddies and waterways, mercury, lead, and dioxins were released. Chen Wu in Guiyu, China, was glad when his village became a dumping ground for discarded computer hardware from the United States and Japan. Salvaging computer parts meant jobs four times the average pay for their rural area of China, even though they poisoned drinking water and created an unsightly landscape of broken circuit boards and hard drives.

They used to deal in pig bones and duck feathers. Now it is integrated circuits, said the head of the Shannon Environmental

Bureau, who gave only his family name, Kuang. Officials in Guiyu would not comment.

Over a period of time working with the parts, Chen Wu's 11 year old daughter grew weak, suffered nosebleeds and was diagnosed with leukemia. Two school classmates were stricken with the same illness. Teachers say more than half the students in school complained of chronic breathing problems.

We did not care much when outsiders talked about the environmental pollution here. "We did not see any harm," said Chen 50, who works at a drug rehabilitation center. "But now our kids are getting sick." Environmental groups consider Gulyu, a cluster of five villages in Guangdong province about 150 miles northeast of Hong Kong, a cautionary tale for countries, or regions that take high-tech waste.

Over the past decade, these groups have taken as much as 80 percent of old computers, monitors, and printers collected for recycling from the Unite States, China, India, and Pakistan, according to a report released by environmental groups. When workers rip through the waste for reusable parts, there are brand names including Compaq, Apple, and IBM. Some components are melted to extract precious metals such as gold and platinum, flat screens, and low grade plastics is burned and the residue dumped beside the rice paddies and water ways. Toxic chemicals such as mercury, lead, and dioxins are released into the air, ground and water.

The first sign of danger was in Gulyu when fish disappeared from a river in the early 1990s, not long after the first truckloads of foreign computer waste rolled in. Chemicals poisoned the wells, so drinking water was trucked in, and the odor of burning plastic is so strong that classes at the nearby Dongyuan Middle School are sometimes halted. The smell is everywhere in the air. One teacher, who gave only his last name, Guo, said about 60 percent of the students and many teachers have trouble breathing. The villagers here are growing richer, he said, but their wealth is built atop the health of other victims.

This year the United States will export as many as 10.2 million discarded computers to Asia, including 9 million to China, the environmental groups report said. The economy has come to depend on computer waste, even though a Chinese government edict banned its importation in 1996. There have been five crackdowns over the past two years, and hundreds of computer waste operations have been shut down. But most were reopened rapidly, often with the help of village officials. The Beijing government has weak control in a region where organized criminal gangs are strong.

There are 2,500 computer waste businesses and most of them are family run. The industry may employ as many as 100,000 people, many of them migrants from other places in China. People in Gulyu have made a living out of waste collection for generations.

It is a full fledged underground economy in this part of China said a man who employs two dozen people stripping desktop PCs from California and Japan. A man identified only by his surname Li, said he buys about 200 tons of computer waste a year from Taiwanese brokers for about $600 per ton. The waste is smuggled via the port of Nanhai and trucked to Gulyu.

Outside Li's dirt-floor workshop, workers used reed baskets to unload a truck full of hard drives, keyboards, and PC parts. Inside, workers rip them apart with hammers and screw drivers. Others sift the debris for anything of value like, tiny nuts and screws, capacitors, and high-grade plastic. In a smaller room, two women held green circuit boards over open coal fires. As the fumes of melting lead solder reddened their unprotected faces and nose. They used pliers to pick off tiny black computer chips. The recovered parts were separated into burlap sacks.

Li said he sells them by weight to buyers, usually from Japan. He makes more than $12,000 a year, 15 times the average rural salary in Guangdong. We are worried about our children, sure, said Li who said he has a 15 year old daughter. But what can I do? This is our livelihood.

Andrew D. Anderson

Chapter 2: What Pollutes Our Waters?

Garbage

Water pollution means lakes and streams choked with algae and weeds, wild life habitat destroyed due to the drainage and filling of wetlands, ground water, beaches closed to swimming, and wells poisoned by toxic chemicals.

Garbage is one thing that contributes to pollution. Years ago when the United States was becoming Industrialized, residents dumped garbage into the streets, factories dumped their chemical's directly into rivers and lakes, cities poured their sewage into the same rivers and lakes, and garbage was deposited into nearby garbage dumps without any concern that it might harm the groundwater.

Ships dumped their garbage overboard into rivers, lakes, the oceans, and postcards showed factories pouring black smoke into the air. America was rapidly becoming industrialized with job opportunities, and most people were proud of the factories in their towns. They knew the sky and the rivers were big enough to absorb anything they might do to them. Even in one part of Manchester, England, there was only one toilet for 212 people.

However, it did create a health problem. New Orleans had typhoid epidemics because sewage was poured into the streets and canals. Memphis lost nearly 10 percent of its population to yellow fever and infant mortality was very high in large cities. In New York, about 120,000 horses pulled carriages, wagons, and streetcars, while Chicago had 83,000 horses doing the same type of work. Every city required large numbers of horses to power the present day transportation. Those horses produced thousands of tons of manure that dirtied the streets and had to be cleaned up and dumped in chosen places adding more pollution.

By the end of the 1800s, city leaders began to recognize that they had to do something about the garbage. They set up garbage

21

collection systems and many cities introduced incinerators to burn some of the garbage. By 1924, about 300 incinerators were operating in the United States and Canada, but many cities began to take their garbage and dump it into the ocean. They did not realize it would only float back to the beaches.

Athens Greek issued the first known edict against throwing garbage into the streets. The state organized the first municipal dumps by mandating that scavengers transport wastes at lest one mile from the walls of the city. This method was not practiced in medieval Europe. Parisians continued to toss trash and sewage out their windows. The west end of London and the west side of Paris became fashionable in the late seventeenth and the eighteenth centuries because the prevailing winds blew west to east, carrying the smell of rotting garbage away from the fashionable section.

State environmental regulators ordered London to stop dumping untreated sewage from its public works building into the Connecticut River. Here in 2002, a dye test by inspectors for the state Department of Environmental Protection found a conduit from a town building sending toilet and other waste directly into the river, according to Anna Symington, a spokeswoman of the DEP. It was not clear how long the sewage had been going into the river.

Problems with Garbage

As the United States became richer and the population continued to grow, the nation produced more garbage and pollution. The Environmental Protection Agency estimates that the average American man, woman, and child threw away about three-fourths of a ton of garbage a year and the amounts of garbage were higher in big cities. Since 1960, America's population grew 34 percent and the amount of garbage produced increased 80 percent.

In the late 1900s, 80 percent of the nation's solid waste was being dumped in landfills, but sites were rapidly filling up and leaking toxic substances into the nation's ground water supply. Not only were many landfills incapable of holding any more trash or garbage, but it did not eliminate the problem. It only created more problems. A landfill is a system of trash and garbage disposal where waste is dumped into a hole and buried between layers of earth.

A lot of states sent their garbage to other states. Some even sent it to other countries, but others countries would no longer accept other garbage. According to the National Solid Wastes Management Association, unless new garbage dumps were built, the nation would run out of places to dump its garbage before the end of last century.

America is a throwaway country today because everything we buy or use has been made to be thrown away whether it was purchased in grocery stores, hardware stores, or drug stores. When the usefulness of anything we have or own is completed, we throw it away.

Ten years ago, The Environmental Protection Agency estimated Americans produced about 207 million tons of garbage or about 4.4 pounds per person per day. In the year of 2000, it estimated America would produce 218 million tons of garbage a year. Most of that was expected to be paper. I signed papers for a mortgage that included over twenty five papers with half of them just for my initials. Ten years ago the garbage paper was 38 percent of all municipal solid waste. Today it is a lot more.

Landfills

Landfills are open areas where cities and towns dumped their garbage. They were once called garbage dumps because cities took their garbage to the area and simply dumped it into a big hole dug in the ground. They were and are the cheapest way to get rid of garbage, but the areas caused more continuous problems. In the United States, almost two-thirds of all municipal solid waste was dumped into landfills. A disposal or tipping fee was charged to use the facilities. The National Solid Wastes Management Association reported that in

1995, the national average tip fee was $32.19 per ton, up 22 percent from the 1992 survey.

When it rained, rainwater dissolved some of the garbage, and it seeped into the ground contaminating the groundwater. The water flowed into lakes and rivers, polluting them and killing fish.

Between 1979 and 1986, the number of landfills dropped from 20 thousand to 13 thousand. In 1995 according to the Waste Age Industry Journal, only 2,893 landfills were still operating. Many landfills were closing only because they were full and not because of polluting of the ground water.

New landfill areas were becoming difficult to find with some states running out of acreage. Connecticut, Massachusetts, New Jersey, and Rhode Island had insufficient area with soil and water condition suitable for landfills. The growing shortage of available landfills was reflected in the 1987 journey of the garbage barge Mobro. It had 3,200 tons of garbage on a barge that floated up and down the Atlantic coast for nearly six months, looking for a place to unload. Rejected by a New York landfill that was full, it was later refused by North Carolina, Florida, Alabama, Mississippi, Louisiana, Mexico, the Bahamas, and Belize. After a 6,000 mile search, the cargo was deposited at the same landfill that initially rejected it.

Experts point out that the newer landfills with multiple liners between the garbage and the ground are supposed to prevent leaks into the ground. According to industry executives, these landfills cost as much as $400,000 and acre to build.

Even though the United States was short on landfills, in 1992 it accepted other countries waste shipped to the United States, primary paper, glass, plastic, and rubber. Canada was the country sending the greatest amount of garbage to the United States and Mexico was the second.

Waste disposal grew to hue proportions. Many sites were filling up and many were leaking toxic substances into the nation's water supplies. America was facing a problem with its ever-growing mountains of garbage. Not only were many landfills in some parts of the country incapable of holding any more garbage, more was learned about garbage and more apparent it was that trucking garbage to landfills did not eliminate it. Municipal governments worldwide were struggling to find the best methods for managing waste. Landfills built and operated prior to the passage of the Resource Conservation and Recovery Act, RCRA, in 1976 were believed to represent the greatest leakage.

In the 1980s used oil was still being dumped in small ponds and swamps next to major and obscure highways. They can still be seen when driving down a highway where growth has not started or is trying to get started.

It is impossible to determine the extent and severity of groundwater pollution in the United States. For years, waste was dumped into open pits or buried in containers that corroded. Cities sprayed icy roads with salts and chemicals, pesticides, and fertilizers were seen as miracles of modern science. We had the mentality that the stronger the chemicals the better the product producing more crops. No one suspected or even thought that much of this material would eventually work their way into the groundwater, only to reappear in our drinking water.

Some of the contributing things were farming, irrigation, septic tank, cesspool discharge, pesticide, landfill, dumps on refuse piles, factories, pumping wells, and sewage leakage.

Each year, 40 million tons of raw sewage, chemical wastes, and other poisons pollute our water supply. The United States Geological Survey figures show that water use has more than doubled in the past 30 years, to about 450 billion gallons per day. Even that demand is small according to officials beside the amount of water available in this country. Stream flow is about 1,200 billion gallons per day and there are estimates of 33 quadrillion to 54 quadrillion gallons in underground water sources called aquifers that should replenished from rainfall and snow-melt with about 300 trillion gallons each year. These figures are questioned later.

Today, as new suburban communities spread out from the cities, the old sites were causing serious contamination problems. In some communities as Dallas, Texas, hazardous wastes were dumped in low-income areas of the city for years. In addition, industries known to produce large amounts of pollutants often were located in such areas.

When a waste site is found to be so badly contaminated with hazardous waste that it represents a serious threat to human health, it is placed on the National Priorities List known as Supervened making it eligible for federal intervention and cleanup assistance. As of 1994, 1,300 landfills were listed on the National Priorities List.

Septic System

Septic tanks and cesspools discharge the largest amount of wastewater into the nation's groundwater. Americans living in rural areas use individual sewage disposal systems, which are the main sources of disease-causing bacteria and nitrates. The largest contamination is septic tanks which discharge waste into the ground each year and greatly increases the contamination in shallow aquifers.

Sub-surface percolation, septic tanks, and cesspools were an estimated 19.5 million domestic on-site disposal systems in the United States in the mid-1970s of which 16.9 million were septic tanks and cesspools. Presumably the remaining 2.6 million systems were privies or chemical toilets. Of all the sources known to contribute to ground water contamination, septic tank systems and cesspools directly discharge the largest volume of wastewater into the sub-surface.

The second largest threat is posed by petroleum exploration and development, and the third is landfills and dumps. The fourth is posed by agricultural chemicals, including dibromochloropane, a pesticide detected in 1993 of 257 wells including 20 public water supply wells tested in 24 California counties. Ethylene dibromide, a suspected carcinogen used as a fumigant for grains, citrus fruit, and soils turned up in 20 percent of the wells tested in Florida. There are an estimated 19.5 million private septic systems in homes in the United States using such facilities, and their total estimated discharge of liquid wastes into the ground exceeds "1 trillion gallons" per year. Until recently, a popular way to clean out septic systems instead of pumping was to use a solvent containing trichloroethylene. It is one of the most common pollutants of groundwater. When pumping out a system, where do you put the waste?

Surface Impoundment

Surface impoundments are pits, lagoons, and ponds that receive treated or untreated waste directly from the discharge point or used to store chemicals for later use to wash ores. Most are small, less than one acre, but some industrial and mining impoundments may be as large as 1,000 acres.

Most are not lined with a synthetic or impermeable natural material, such as clay, to prevent liquids from leaching into the ground. Eighty seven percent of the impoundment's are located over aquifers currently used as sources of drinking water, while less than 2 percent are located in areas where there is no groundwater or it is to salty for use. About 70 percent of the sites were located over very permeable aquifers that would allow any contaminant entering its waters to spread rapidly. Researchers found that groundwater protection was rarely if ever considered during the site selection of impoundments.

Human wastes disposal activities are, industrial waste impoundment's and landfill, municipal landfill operations, underground injection operations, and individual septic tank systems. As of 1982, Environment Protection Agency located over 180,000 waste impoundments at some 180,000 sites. Of the industrial sites evaluated in the agency's 1980 Surface Impoundment Assessment, 70 percent were found to be unlined, 50 percent were sitting directly on top of aquifers used as sources of drinking water, and 98 percent were located within one mile of a water supply well. The Environment Protection Agency believes that there are 75,000 operating industrial landfills containing non-hazardous waste and about 200 hazardous waste landfills.

Rubber Tires

An estimated 242 million tires are discarded annually with 80 percent from passenger cars and the remainder from trucks. Over 50 percent of the rubber used in the United States is used to make tires and approximately 7 percent of tires are recycled. The remainders are stockpiled, illegally dumped, or incinerated. Two to 4 billion tires have accumulated in landfills or uncontrolled tire dumps that pose health and fire hazards. They are highly combustible, do not compost, and degrade slowly but not complete. The material, primarily hydrocarbons, burns easily producing noxious air pollutants and toxic runoff. The tires do not compress in landfills and provide breeding grounds for a verity of pests. Some states ban the disposal of tires in landfill.

There may be a savior. The tire is not like paper which can be recycled into pulp, plastic, metals, or glass which can be melted. Vulcanized rubber has resisted recycling. The tires discarded yearly weights more than three times the 1.7 million tons of debris removed from the World Trade Center wreckage. The Rubber Manufacturers Association which is an industry trade group, estimates 10 percent to 15 percent still wind up in dumps each year creating fire hazards and breeding grounds for rats and mosquitoes.

Tire management uses them as fodder for road builders, a fuel for power stations, and playground equipment. But there is a company named Farris and other scientists in the country that are developing techniques to remold old rubber products into entirely new ones. Avram Isayev, a professor in polymer engineering at the University of Ohio at Akron, has discovered a way of using ultrasound waves to break rubber's molecular bonds. A former graduate student named Jeremy Morin discovered that he could recycle rubber by breaking it into tiny crumbs and subjecting them to extreme heat and pressure making it good enough for shoe soles and windshield wipers.

Tires can generate 24 percent more energy per pound than coal and they burn cleanly in a controlled incinerator different than fires at tire dumps that send up black smoke for a long period of time with odors. Exeter Energy in Sterling, Connecticut burns 30,000 tires a day making it one of the world's biggest tire-to-energy plants and the emissions are below state limits.

Forty-two percent of the nation's tires are burned each year, but many are ground into crumb rubber and used in civil engineering projects, where they serve as filler in concrete and asphalt. Also tires are reused as ship bumpers, artificial coral reefs, and paint additives.

The problem with recycling rubber tires is that natural rubber is a polymer, or a collection of thousands of repeating molecules bonded together to form long chains. The results is small molecules and polymers, like rubber, are more like a bowl of spaghetti said William's in the lab in Akron, Ohio. It is also like pasta. Untreated rubber is as flimsy as a cooked noodle. Mr. Charles Goodyear discovered that in 1839 when he accidentally dropped a mixture of India rubber and sulfur onto a hot stove, it turned a flimsy gum into the sturdy stuff of modern tires. It fused the rubber molecule chains together creating something flexible but almost indestructible.

The heat and pressure from vulcanization binds the chains together with a cross linking agent that is usually sulfur. The crosslink's are what make rubber so difficult to recycle because the carbon-sulfur bonds prevent the rubber from melting or chemically breaking down. Some scientists at U Mass found ways to circumvent the problems of vulcanization called high-pressure, high temperature sintering. By using tens of thousands of pounds per square inch of pressure to squeeze the voids out of crumb rubber and put the particles into intimate contact. Hundreds degrees of heat then catalyzes a series of reactions that break and reform the rubber bonds resulting in a new material that can be molded into any shape.

Avram Isayev at the University of Ohio is perfecting a technique based on ultrasound which is sound waves above the frequency of human hearing. In medicine, ultrasound bounces off body parts to create pictures. In the ocean, whales use ultrasonic echo-location as a form of sonar, but the sound waves of energy could also de-vulcanize rubber by preferentially breaking apart the carbon-sulfur cross-links which are its weakest bonds. Once broken down, the rubber can be treated like virgin rubber and re-vulcanized into new shapes thus the many tires discarded may be useful in the future and cause less pollution when their useful life is over.

The Worlds Largest Garbage Dump

Methane, a flammable gas, is produced when organic matter decomposes in the absence of oxygen. If not properly vented or controlled, it can cause explosions and underground fires that smolder for years. This gas is being recovered through pipes inserted into landfills and distributed or used to generate energy. Less than one third of the nation's toxic waste dumps meet requirements for monitoring underground water supplies near their sites under current disposal laws.

The most ominous example of the nation's garbage problem sits on New York's Staten Island. It is listed in the Guinness Book of World Records as the largest anywhere. The 50 year-OLE Fresh Kills landfill is about as big as the largest of the Egyptian pyramids in heights and volume. When it closes some time after the turn of the century, it will be more than 500 feet tall and will rival the Great Wall of China as the largest man made structure in the world. It opened in 1948 on swampy lowland and is the tallest mountain on the Atlantic Coast between Florida and Maine. When it closes, the city plans to turn it into a grass-covered park.

Garbage usually accumulates in four ways:

1.   Garbage may be dumped into a landfill or garbage dump.
2.   Garbage may be dumped into a surface impoundment which is a large pond in which it can be stored and then treated so it can be safely thrown away.
3.   Waste can be spread out in a "land application" where it is taken to an area and spread on the land.
4.   It may be dumped onto a "waste pile" on the ground where it is stored and it may be treated.

All four methods are unacceptable. That is the responsibility of each state to count its own dumps and report on its compliance with federal pollution standards. In 1993 new standards went into effect for landfills. The EPA estimated that half of the 6,000 United States landfills would close within another three years, at least in part because of non-compliance with the new rules. They were spread all over the country.

New England is the only region that landfills less than 50 percent of its waste. It also incinerates more garbage than other regions. Municipal Solid Waste incineration occurs almost entirely east of the Mississippi.

How Garbage Decomposes

In 1992, the Smithsonian Institution reported on its Garbage Project, an archeological study prepared at the University of Arizona on 14 landfills. The findings were that although some degradation takes place initially sufficient to produce large amounts of methane and other gases, it then slows to a virtual standstill. Methane is a colorless and odorless flammable hydrocarbon. The study reported that the overall volume of old organic matter recovered largely intact turned out to be astonishingly high. Even after two decades, one third to one half of supposedly degradable organic remained in recognizable condition. The Smithsonian concluded that well-

31

designed and well-managed landfills seem more apt to preserve their contents for posterity than to transform them into humus or mulch.

The aerobic stage is living, active, or occurring only in the presence of oxygen. Garbage put in a landfill begins to decompose with the help of oxygen. Aerobic bacteria reduce water, carbon dioxide, nitrates, partially degraded organic material, and produce heat, often 122-158 F. This stage lasts about 2 weeks, until the oxygen is deleted.

The acid anaerobic stage is after the oxygen is gone and garbage continues to decompose. Anaerobic bacteria produces carbon dioxide and partially degraded organic material and organic acids. Heat production is reduced. This stage lasts 1 to 2 years.

The methanogenic anaerobic stage or methane gas is formed as a product of anaerobic decomposition. Methane and carbon dioxide are the dominant chemicals produced. This stage can last several years or decades depending upon landfill conditions, temperature, soil permeability, and water. Methane gas can be recovered during this stage.

Even among the most advanced landfills, some will eventually leak. Methane, a flammable gas, is produced when organic matter decomposes in the absence of oxygen. If not properly vented or controlled, it can cause explosions and underground fires that smolder for years. Methane is also deadly to breathe. Increasingly, the gas is being recovered through pipes inserted into landfills and distributed or used to generate energy. The Smithsonian Garbage Project found that 15 to 20 years after a landfill had stopped accepting garbage, the wells vented methane in fairly substantial amounts. In time, methane production drops off rapidly indicating that the landfill has stabilized.

Military Bases

The military is guilty of a lot of military sites contaminating our drinking water. Cape Cod has those water problems. A chemical used in military explosives has contaminated the drinking water for thousands of people in the town of Bourne, providing the first clear evidence that pollution from years of training with grenades and rockets at the Massachusetts Military Reservation is now seeping into Cape Cod's underground water supply.

Three of the Bourne Water District's six wells have been shut down in the last month of 2002 after trace levels of per chlorate were discovered. Officials anticipate a fourth well will be closed soon because the chemical is moving through nearby ground water at the rate of about a foot a day. The four wells clustered together provide 70 percent of the water for a summer time population of 19,000 people. Also a well at a private Sandwich home was found to contain the same chemical.

This is our worst nightmare come true, said Tom Cambareri, water resource program manager for the Cape Cod Commission, a regional planning agency. "Not only are the contaminants being found in the ground water, but in public water supply wells, and it is migrating off the base."

With the region already hit hard by a drought, Bourne and state officials are now scrambling to find enough water for residents by summer time. Bourne Water District manager Ralph Marks flew to Washington to ask the Army for money to build a 3-mile pipeline to reach a water supply on the northeast corner of the base. Bourne officials may also be able to get water from nearby towns, but even if that succeeds, residents may be in for severe water restrictions this summer.

Environmental and military officials have been fighting over the use of explosives for National Guard training at the base since 1997, when the United States Environmental Protection Agency ordered a suspension of live ammunition until its environmental effects were better understood. At the time, military officials insisted that explosives had not contaminated the ground water.

Endangered birds find a home at the polluted base in Sandwich, Massachusetts Military Reservation. It is not a very pretty place, marked by huge swaths of bald ground, plenty of paved roads, and squat drab buildings. It is one of the largest, most complex, and expensive Superfund sites in the United States. The Cape Cod military base is blamed for contaminating ground water, killing off cranberry bogs, and raising blood pressures in surrounding neighborhoods with stray and sometimes un-detonated ordnance.

For some unlikely inhabitants, this place is 20,000 acres of absolute heaven. Upland sandpipers, birds that are on the endangered species list, love the place, as do grasshopper sparrows and vesper sparrows. Both threatened species, together with a host of other delicate feathered creatures. Last year, 1998, nine upland sandpipers and 10 grasshopper sparrows were spotted on the reservation.

The reservation includes Camp Edward's of the United States Army, the Otis Air National Guard Base, the Coast Guard Air Station, and the Cape Cod Air Force Station. The fact that it is also home to such fragile species is not lost on environmentalists. With all of the pollution and toxic ground water, that part of the base provided a habitat for an endangered species, said Andrea Jones, grassland conservation coordinator with the Massachusetts Auburn Society. The sandpiper needs at least 100 acres of grass to be happy, and that grass should be long enough to nest in but short enough to see predators.

Newly released Environmental Protection Agency (EPA) documents released in 2002 said there are 16,000 inactive military ranges including biological weapons and chemical weapons that pose an "imminent and large public health risks." It requires the largest environmental cleanup program ever implemented by the US government, excluding the nuclear waste cleanups.

A Washington-based advocacy group called the Public Employees for Environmental Responsibility says the EPA officials are concerned by the Pentagon's refusal to abide by EPA regulations when cleaning up the sites. One document cites a disturbing trend by the military services and the Army Corps of Engineers to limit their cleanup activities and take ill-advised short-cuts to limit costs. It is feared there will be much greater and costlier cleanup problems associated with unexploded ordnance than previously acknowledged by government officials. Once again they are talking about the cost rather than if the problems can be completely eliminate for a one hundred percent cleanup.

Raymond F. Dubois is deputy undersecretary of defense for installation and environment, agrees that the cleaning up of unexploded ordnance could cost any where from $14 billion to several times that much, "depending on the eventual use of the land" or how complete the cleanup is. Nothing was mentioned about the time it will take to clean up 16,000 inactive military ranges on 15 million acres of former military land and 25 million acres still in the Pentagon's possession as well as the large number of nuclear plants and the nuclear storage facilities that are leaking.

Jeff Ruch is the executive director of the advocacy group and believes the EPA and Defense Department are failing to adequately address ground water and soil contamination caused by unexploded munitions on inactive ranges across 30 million to 40 million acres. That is equivalent to the size of the state of Florida.

The cleanup of unexploded ordnances on military ranges will be the largest environmental cleanup program ever in the United States and probably the world. Some of the ranges cover 100 to 500 square miles and most of the properties are formerly used defense sites where the military has relinquished control and are now being utilized for purposes other than a military range yet humans are in contact with the area not knowing the potentials of life. One interim report said out of nine explosions of military munitions that had been cleaned up, six involved fatalities. Were the humans told of the potential possibilities?

Andrew D. Anderson

Farms Cause Pollution

Banks required farmers to use pesticides to qualify for crop loans and pesticide companies spend heavily on advertising and lobbying. Government subsidies and pricing policies encouraged pesticide use, but in underdeveloped countries, funds and expertise were lacking.

Some supporters of pesticide and fertilizer use contended that low levels of chemicals in the ground water should be acceptable, and pesticide supporters had a great deal of influence in Congress and had successfully halted much groundwater legislation.

Chemical fertilizers contributed to the increase in the world's food production. Its use and food production both increased at a rate of 7 percent per year from 1950 to 1984. Later in 1984, that increase dropped to less than 2 percent annually. As the world added 90 million people per year, the need for food supplies expanded. Yet, the growth in world fertilizer use actually slowed.

In order to satisfy themselves, farmers said pesticides evaporated or degraded into harmless substances. Now they wonder if it poisoned their drinking water. It was later determined that pesticides spread into the air, water, and food chains and American farmers slowed in using the pesticides, but foreign countries increased. It is also not safe for wild life.

It is a no win. The development of an insect that is resistant to a pesticide is virtually automatic, because those insects that survive are unaffected by that chemical and will breed off in the spring insects that are also immune, creating super-bugs. At first use insecticides reduced crop losses, but over time, the pests rebound. World wide, the number of pests resistant to chemicals continued to climb. In 1938 scientists knew of only seven insect species resistant to pesticides. Today, many hundreds of pests, including most of the world's major pests, are immune to existing pesticide chemicals.

36

Farmers have found that the little extra fertilizer that they once used is lost in nutrient runoff, which creates stream pollution in agricultural areas. The major factor is the decreasing response of crops to fertilizer. During the 1960s, an additional ton of fertilizer applied on a farm in the United States corn-belt boosted output by 20 tons. Today, another ton may boost only a few tons. The Drops are apparently approaching the limits of photo synthetic efficiency.

Using Sludge As Fertilizer

It is a muddy precipitated solid matter produced by water and sewage treatment processes with remaining nutrients that is not only used where factory farms are but in areas where families live. A family living in Minify, California says treated sewage sludge as fertilizer was dumped five out of seven days on the adjoining property and smelled like dead bodies. Their 5 year-old daughter experienced chronic sinusitis and bacterial infections, and the husband ended up in the hospital with a staff infection. After the wife was treated for nose and throat infections, her doctor recommended they move. They had been breathing the air from farms near their Minify, California home where the sludge was used.

Well over half of the 5.6 million tons of sewage sludge generated each year in this country gets recycled into our soil which is also known as bio-solids. It is rich enough in nutrients to be used as fertilizer, and "toxic" enough to warrant regulation. The Environmental Protection Agency enacted Part 503 of the Clean Water Act in 1993 to regulate sludge, but it has been questioned to the strength and efficacy of those rules. The EPA says little scientific evidence shows a link between sludge and illness.

The EPA analyzed data on 411 pollutants in sewage sludge. It set limits on how much arsenic mercury, cadmium, and other inorganic chemicals could be spread on farmland. The EPA knew sludge contained other pollutants and bacterial pathogens, such as salmonella, E. coli, endo-toxins, asbestos, and viruses that cause polio, hepatitis A, and meningitis, but decided to rely on sludge treatment and its exposure to air, heat, and sunlight to minimize their harm.

Now nearly everything is flushed down toilets and poured into drains by industrial plants, hospital, gas stations, and householders end up at waste-water treatment plants. The water with toxins removed, is treated and returned to waterways. There is a class B sludge that contains more pathogens, but is a much more nutrient-rich fertilizer and is used as fertilizer. In some cases farmers get it free to use and some even get paid to take it. It is also spread on golf courses, parks, and lawns.

A theory developed by the Department of Agriculture and the EPA held that toxic chemicals and pesticides leaching out of sludge fertilizer would be captured by clay-bearing soil below it indefinitely. What if there was no clay in the ground? However Mr. Robert Swank Jr., a former EPA research director in a September 6, 2000 report said we know that is not true.

The EPA's rule controls when sludge should be tilled into the soil. It limits public access, the planting of crops, and the grazing of cattle. Buffer zones must separate sludge-treated land from slopes, creeks, and homes. To track compliance, waste water treatment plants and sludge hauling companies must keep records on all this for the EPA's review. But in 1998, the EPA's inspector general found little more than a third of those reports were reviewed by the agency's small bio-solids staff.

Two years later the EPA found the agency had reduced its regional bio-solids review staff by 44 percent and there were two people for eight states. Virginia's Supreme Court ruled in January 2001 that local governments cannot stop land owners from using sludge as fertilizer. Other states say they do not have the resources to effectively police sludge use.

Cornell Waste Management Institute has tracked more than 250 sludge-exposure complaints in at least 25 communities. Some people believe they have contacted illnesses from breathing the sludge dust from nearby fields. Others believed they have been infected by direct contact with soil carried by storm runoff.

There are also lawsuits that have been filed. A New Hampshire family claimed that 26 year-old Shane Conner died in 1995 from a respiratory illness caused by winds that carried from a nearby farm fertilized with sludge. The family received a settlement earlier this year from Synagro the sludge-hauling company it sued. The family of 11 year-old Tony Begun believes his 1994 death was the direct result of his biking excursion through an unmarked field spread with sludge.

In Georgia, two dairy farms are suing Augusta County for negligence in treating and spreading sewage sludge. Cattle at both farms experienced high mortality rates and Illnesses after grazing on fertilized fields.

What We Throw Away

The United States generates more garbage than any other nation on earth and twice as much per person as the European Countries. The environmental Protection Agency estimates that Americans produce about 1,400 pounds of trash per person per year, more than four pounds per day per person. In addition, American industry is responsible for nearly 270 million tons of toxic, corrosive, or ignitable waste annually.

Garbage generation expands as cities grow. Consumers want more convenience foods, with their packaging such as disposable diapers. Before the days of densely populated urban areas, waste disposal was eased by the ability of the surrounding land and water to absorb the wastes. A century ago, farm communities created little waste. Today farm communities produce waste similar to urban areas in addition to runoff from insecticides and fertilizers used in the fields.

Andrew D. Anderson

An inventory of what Americans throw away would reveal valuable metals, paper representing millions of acres of trees, and plastics incorporating highly refined petrochemicals. It is indicative of American priorities and values that products so rich in raw materials are frequently considered worthless. The contamination of the air, water, and soil with hazardous wastes can cause serious health effects, and new compounds are constantly being developed.

Most garbage was paper and in 1993, 38 percent of all municipal solid waste was paper followed by yard wastes, 16 percent plastics, 9.3 percent rubber, leather, textiles, 9 percent metals, 8.3 percent food wastes, 6.7 percent glass, and 6.7 percent wood with a high probability of ground water pollution.

Nearly $1 out of every $12 Americans spend for food and beverage pays for packaging. According to the United States Department of Agriculture, packaging is the second largest portion of the cost of food since marketing is the largest portion. The increasing numbers of women in the workforce and changes in family structure have resulted in greater demands for convenience products, carry out meals and frozen, and vacuum packed foods. Soft drink consumption has risen with its increase in cans and plastic containers.

Aluminum, the most abundant metal on earth, is never found free in nature. Scientists first refined the metal into a valuable product in the 1820s. Its use has continually escalated. Today beverage cans are the largest single use of aluminum and much waste. More than 46 types of plastics are produced in the United States which is more than aluminum and all other nonferrous, containing no iron or metals combined. Most of these plastics are non-biodegradable and, once discarded, they remain relatively intact for many years.

Chapter 3: The Major Roles Ground Water Plays

Ground Water

Each year, 40 million tons of raw sewage, chemical wastes, and other poisons pollute our water supply. United States Geological Survey figures show that water use has more than doubled in the past 30 years to about 450 billion gallons per day.

Fertilizer use is associated with mass production-large scale farming. The amount of fertilizers used increased from 31.8 million tons in 1965 to 47.6 million tons in 1978. Ground water contamination by fertilizers results from that crop production uses of only fifty percent of the nitrogen, twenty percent of the phosphates, and thirty-five percent of the potassium applied as fertilizers. The rest leaches into the soil, is lost in soil erosion, or is part of runoff leading to eutrophication of waters.

This leads to water pollution. Afflicting surface and groundwater, pollution has severely limited the amount of drinking water available. Hazardous wastes have leached out of landfills, animal waste, fertilizers, herbicides, and pesticides have contaminated much agricultural land.

We now recognize that a good deal of the precious drinking water supply of this country in underground aquifers has been put in jeopardy by irresponsible disposal practices, and we are beginning to understand the enormous threat presented by the dumping of unprotected toxic wastes in unprotected landfills, said United States Rep. Albert A, Gore Jr., Democrat of TN during a hearing before a House Science and Technology subcommittee.

In New Jersey, leakage of chromium, copper, zinc and other toxic metals and chemicals from industrial lagoons have contaminated millions of gallons of groundwater. New Jersey's Department of Environmental Protection, part of a study conducted for the National Cancer Institute, surveyed public wells serving a large number of customers. "The results of these tests are sobering," said Thomas Burke, director of the study. The majority of the drinking waters contained low levels of potential cancer-causing volatile organics. These results, he continued, clearly indicate the sensitivity of our water resources to the threat of chemical contamination. Evidence of toxic contamination was found in urban, industrialized areas, as well as the most rural parts of the state.

Industrial waste is not the greatest threat to drinking water. That position is reserved for septic tanks, which are estimated to exude some 800 billion gallons of effluent into United States soils annually. The second greatest threat is posed by petroleum exploration and development, and the third greatest is landfills and dumps. The fourth-ranked threat is posed by agricultural chemicals, including dibromochioropane, a pesticide detected in 193 of 257 wells, including 20 public water supply wells tested in 24 California counties. Ethylene dibromide, a suspected carcinogen used as a fumigant for grains, citrus fruit and soils turned up in 20 percent of the wells tested in Los Angeles California.

Other contaminants include seawater and acid rain. On Cape Cod in Massachusetts, some towns must post weekly warnings of the salt content in well water, because seawater is creeping into aquifers as the groundwater is slowly being exhausted. Acid rain has become so bad in some northeastern states that all fish and other animal's life in the lakes have been killed, and airplanes dump lime on the ice during the winter to neutralize the water once melt occurs.

In 1950, the United States took 12 trillion gallons of water from the ground. By 1980 the figure more than doubled and is still going up. Twenty one billion gallons of water flow in from rain, snow melt, and water return. The fertile crescent of the Middle East, now semi-desert, was heavily irrigated in Biblical times. During the rule of the Pharaohs and Ptolemies, water of the Nile was carried to large areas that now are little more than desert. Today the American Southwest is the most heavily irrigated areas in the world, transforming a desert into a veritable Garden of Eden.

In the desert areas of the Southwest, the heavy irrigation may give way gradually to greenhouse farming, which uses less water and produces higher yields. New growth forests and orchards could protect soil from drying out and act as rain forests. Saving water for Americans to drink is a greater threat to American well-being and prosperity than any external force. It deserves priority action by Administration and Congress, but I do not think they realize it.

Where does groundwater come from? Groundwater is within the earth that supplies wells and springs. It lies under the earth's surface almost everywhere, beneath hills, mountains, plains, and even deserts. An aquifer is a water bearing stratum of permeable rock, sand, or gravel and plays a very important part in the supply of groundwater, drinking water, and water for irrigation. It may be close to the surface or many hundreds of feet below the surface, as in some very dry areas of the American West.

This water fills the openings or cracks in underground rocks and may have been under the earth for hundred or even thousands of years. Some rain water or snow evaporates into the atmosphere, some runs off into streams, and some seeps into the soil and openings and cracks in the rocks underneath the ground creating groundwater.

The Ogallala Aquifer is a large underground reservoir of water that transformed much of the Great Plains into one of the richest agricultural areas of the world, but is now being sucked dry.

It supplies groundwater to one fifth of all the croplands in the United States. Natural replenishment should raise the water level in the aquifer about one half of an inch every year, but pumping water out for irrigation lowers it four to six feet every year. By the late 1980s the aquifer had been half emptied. No one knows for certain when the aquifer will run dry, but when it does, the effect on American agriculture will be very bad. Aquifers sometime collapse after they are drained and the collapse of the underground chambers not only cause sinkholes and settling in the land above, but also means that the aquifer can never be refilled.

Agriculture is the largest user of groundwater. Forty percent of water used by agriculture goes for irrigation, 55 percent consumed by livestock, and it supplies 30 percent of water used by industry. In the East and South, groundwater is used primarily for industrial and domestic purposes. In the West, 88 percent of ground water is withdrawn for crop irrigation.

Groundwater reservoirs, or aquifers, contain nearly 50 times the volume of the nation's surface waters, constitute 96 percent of all the fresh water in the United States, and they are primary drinking water sources for half of the population, nearly 115 million people.

Increasing concern is being expressed because groundwater in some locations is contaminated by toxic or potentially hazardous chemicals, many of which are known or suspected carcinogens.

The Ogallala Aquifer

Three years ago the newspaper Times reported, "The Ogallala Aquifer, the vast underground reservoir of water that transformed much of the Great Plains into one of the richest agricultural areas of the world, is being sucked dry."

The aquifer stretches from South Dakota through Nebraska, where two-thirds of its water lies, to Wyoming, Colorado, Kansas, Oklahoma, New Mexico, and Texas. For the past three decades, farmers have pumped water out of the Ogallala as if it was inexhaustible. They now disperse it prodigally through huge center-pivot irrigation sprinklers, which moisten circular swaths a quarter-mile in diameter. The annual overdraft or the amount of water not replenished is nearly equal to the flow of the Colorado River.

A report by a Boston engineering firm company, Dresser & McKee, estimates that by the year 2020 some 5.1 million acres of irrigated land will dry up. Some believe the report is too optimistic. Nearly 12 percent of our cotton, corn, grain, sorghum, and wheat are watered by the Ogallala. Almost half the nation's beef cattle are fattened on high plain feedlots. In Texas alone, 70,000 water wells have been dug into the aquifer. Parts of the Panhandle have already used up more than half the water in the portion of the aquifer beneath them. Farm manager Jim Bell admits, "We know we're losing our water. We've just got to learn to use it less and better."

Los Angeles and San Diego is still the growing megalopolises of California and the rich San Joaquin Valley that grows everything from oranges to cotton must import water from distances of hundreds of miles. But that water may be reduced this year in 2002 because of a Supreme Court decision, turning more water from the Colorado River to Arizona.

The large underground reservoirs were filled by natural processes over many thousands of years. Today they are being emptied at a fast rate without their being even concern. They have always been there for us.

A well in Gaza Strip, between Israel and Egypt, draws water from the Gaza Aquifer. Demands on the wells in the crowded strip have depleted the aquifer by more than 50 percent. And the water from the wells is getting saltier and saltier because as the level of fresh water in the aquifer drops, salt water seeps in from the nearby Mediterranean Sea. The aquifer is not irreversibly contaminated yet, but such contamination is common when aquifers near bodies of salt water are allowed to become depleted.

Non-Point Water

Non-point sources are currently responsible for about three-fourths of the sediment in our rivers. Over 90 percent of fecal and other bacteria, 80 percent of the total nitrogen, and 50 percent of the phosphorus reach our waterways. They are the major sources of pesticide pollution, and an important source of other toxic pollutants. Over 90 percent of the lead reaching most of the Great Lakes comes from non-point sources.

Non-point source pollutants from agricultural activities and urban areas cause the majority of non-point problems in our nation. The Missouri basin, with its high density of livestock operations, has an organic pollution waste-load equivalent to the population of China. Some of these operations are classified as point sources and many are classified as non-point sources.

Excessive loading of nutrient from animal waste and cropland have caused serious eutrophication previously mentioned, in lakes, alga blooms, weeds, and fish kills that concern the public.

Non-point source pollution provides approximately one-half of all the conventional water pollution problems in America. In many cases, non-point source contributions from agriculture, urban areas, forestry, and construction sites overwhelm the efforts made to date to address the point source contributors.

Non-point pollution is now a problem in virtually every bay and estuary in the United States, and it will get worse. It is made more acute by coastal development, the process of ditching, draining, and filling wetlands that previously served both as breeding ground for marine life and as filters for pollution from the land. At least half the nation's coastal wetlands have now been lost to development, the EPA's Rebecca Hammer says.

The long term impact on fish, shrimp, crabs and other food can only be estimated, but pollution from nitrogen and phosphorous compounds are already altering the balance of nature and is some of the nation's most productive fishing grounds, Albermarle and Pamlico Sounds in North Carolina are two cases in point. Veteran crabbers who have been fishing the Pamlico River for twenty years say their catches have dropped by half in the past two years because the river's oxygen content has been depleted. Paul says, "Pamlico Rivers blue crabs are coming ashore in a desperate search for oxygen. Last night they were even climbing on top of the pier because of the dead water."

The burden of non-point pollution is demonstrated all too well in Chesapeake Bay, one of the mid-Atlantic regions' busiest waterways. Like other endangered estuaries, the Chesapeake has clearly benefited from federal and state controls on point-source pollution from sewage and industrial plants. But fertilizer and pesticide runoffs from upstream farms are still a major problem. According to a recent study, it probably comes from acid rain, a finding that may have enormous implication for other coastal waters and for the continuing campaign to pass acid-rain legislation.

Millions of dollars spent to clean up point sources are being overwhelmed by non-point source pollution. Seventy five percent of all of our sediment and 90 percent of phosphorus is a major cause of pesticide pollution. The pesticides will stay in the soil and largely break down were it not for loss of soils to waterways and then to converting into serious pesticide pollution problems. Also, our non-point source problem is a major cause of sources of biological oxygen demand.

Sewage Disposal

Two decades after the Clean Water Act set a national goal of returning all waters to fishable and swim conditions, sewage remains one of the nations' greatest problems. Treatment of sewage is inconsistent, even within a given sewage system. While most sewage passes through a first stage which skims off grit and heavy material, most sewage does not reach the second stage where it would be aerated, chlorinated, and further decomposed. More than a quarter of all United States sewage flows into waterways without treatment by disinfectants or filters.

Most of the raw sewage is body waste, which mixes with industrial wastes and chemicals, curbside runoff, and storm-water drainage. Millions of gallons of raw sewage wash through the sewer system into the ocean or other waterways. Some of it will be treated at sewage plants and turned into what is called "sludge" a watery black mud that is supposed to be aerated, baked, and decomposed.

This sludge must then be deposited in landfills, incinerated, dumps, spread on farm land with restrictions or disposed of in some manner.

Any dumping ground landfills, incinerators, or waterways will become the repository of nutrient overload, toxic polluters, and non-degradable matter that will end up somewhere in the environment. Traditionally, the responsibility for dealing with waste water has fallen to communities, but growing populations and increasingly complex environmental problems have created the need to centralized regulation. The clean Water Act was legislated in an effort to resolve these problems.

Researchers are studying alternative methods of sewage disposal, including using bio-purification with micro-organisms and plants, recycling methane to heat greenhouses, and applying treated sludge as fertilizer on farmlands.

Pesticides and Fertilizer

Cleaning up the nation's groundwater is expensive. The costs for cleaning up typical groundwater contamination from a chemical landfill are estimated at $5 to $10 million per site. If the pollution is localized, it may be more practical to simply shut down the contaminated wells and find water elsewhere. Cleanup options range from capping a section of an aquifer with a layer of clay that prevents more pollution, to more complex and expensive methods such as pumping out and treating the water and then returning it to the aquifer which is not always possible.

To control dangerous chemicals used on farms, the law requires the Environmental Protection Agency to register the pesticides farmers use against insects, rats, mice, etc.

Their use continues to grow. Unlike most chemicals, pesticides are designed to kill or alter living organisms of anything living. According to the United Nations World Health Organization, pesticide poisonings affect about 2 million people per year. Of those, three quarters suffer health problems such as dermatitis, nervous systems disorders, and cancer. The organization estimates that 4,000 to 19,000 people die annually as a result of pesticides.

Many experts believe this is only the beginning, and evidence of pesticides in the air, water, and food chains are being found.

Drinking water in Des Moines, Iowa, was so contaminated by agricultural chemicals that the local nightly news announced tap-water nitrate levels. The best of the new pesticides can be used in minute quantities and show few signs of health problems, but they are more expensive. What are minute quantities? Many farmers continue to use the older pesticides, especially in developing countries. DDT, a colorless odorless water-insoluble crystalline insecticide that tends to accumulate in ecosystems and has toxic effects on many vertebrates, is now banned in most industrialized countries. It is still used in developing nations where many farmers see it as an inexpensive way to control pests. Even the new formulations are not entirely safe, especially for wildlife. The United States Fish and Wildlife Service reports that pesticide harm about 20 percent of the country's 681 endangered species. They create the need for more and different chemicals.

The development of a resistant strain of insect to a pesticide is virtually automatic, because those insects that survive are unaffected by that chemical and will breed offspring that are also immune. In the United States overall crop losses due to insects have actually gone up since DDT's discovery. At first, insecticides caused pests losses to increase, but over time, the pests rebounded and a new pesticide was found to destroy immune insects. Worldwide, the number of pests resistant to chemicals continues to climb. In 1938, scientists knew of only seven insect species resistant to pesticides.

As an alternative to pesticide use and abuse, many scientists are encouraging the practice of Integrated Pest Management, a system that combines biological control, like natural predators of pests, cultural practices, planting rotation, diversification, genetic manipulation, pest-resistant, and crop varieties with a modest use of chemicals. Rather than attempting the impossible task of eliminating pests, the goal is to strike a sustainable profitable balance with nature.

Some banks require farmers to use pesticides to qualify for crop loans. Second, pesticide companies spend heavily on advertising and lobbying. Third, in underdeveloped countries, funds and expertise are lacking. Fourth, IPM is something new and different, and old habits and beliefs are hard to break. Finally, government subsidies and pricing policies encourage pesticide use. It is estimated that United States farmers could cut their pesticide use by up to 50 percent and still not lower harvests or significantly raise costs.

Organic pesticides and fertilizers, which come from biological sources rather than chemical factories, are beginning to compete successfully in the marketplace. Industry reports indicate that organic materials made up about 5 percent of the market in 1995. These products include pyrethrum and enema, extracts of flower and tree seeds, which cause no contamination from handling or from leaching into the water supply of food chains.

Some critics of organic pesticides and fertilizers contend that low levels of chemicals in groundwater should be acceptable. They also argue that some areas should be designated for agribusiness and allowed to become polluted. They emphasize the uncertainty of the health-risk calculations, especially since tests have been done not on humans, but on animals. Pesticide supporters have a great deal of influence in Congress and have successfully halted groundwater legislation.

Benzoate, a fungicide once considered the farmer's best friend, has become his or her worst enemy, causing plants to wither and die. For two decades, benzoate was the farmer's secret to growing unblemished fruits and ornamental plants. Some farmers in Florida now find their land is unusable for farming. Researchers think the once helpful chemical, when combined with humidity, heat, and other chemicals, became ponies over time. The EPA banned parathion use in 1991 after deaths and illnesses among California farm workers. Wildlife experts in California reported that a pesticide spill from a train derailment, "killed every living thing in the Sacramento River and Lake Shasta and that it will be years before the fishery is restored."

In 1982, the chemical dioxin became Toxic Enemy Number One. In Times Beach, Missouri, the soil was declared contaminated with dioxin after a hazardous waste trucking company sprayed roads and parking lots, for dust control, with waste oil containing dioxin in the early 1970s. The government had to permanently evacuate all 2,240 residents and the state and federal governments bought the property of residents for $37 million. About 100,000 cubic yards of contaminated soil was removed, awaiting the construction of an incinerator at Times Beach.

It comes in 75 varieties, including the defoliant Agent Orange, which some Vietnam veterans blame for their cancers, nervous system disorders, and genetic defects in their children. Many of those soldiers have been granted disability benefits by the federal government, although the connection between their problems and dioxin has not been scientifically proved. Now environmentalists have expressed concern over the emission of dioxin from incinerators. Incineration is identified by the EPA as the primary cause of known dioxin emission.

Use of Ground Water

Approximately 53 percent of the nation's population depends on groundwater as the primary source of drinking water. The percentage is much higher in rural areas, where more than 90 percent of the population depends on groundwater. Its use increases faster than the rate of population growth.

Between 1950 and 1980, worldwide water use more than tripled. In the United States it increased by 150 percent during that 30 year period, although the country's population grew by only 50 percent. In 1990 the United States used about 400 billion gallons of water each day, more than any other industrialized nation. A typical

American family of four people used 243 gallons each day, and three-quarters of this amount was used in the bathroom.

People use water in five basic ways domestically, for drinking, cooking, bathing, washing, and sanitation. In industry, it is used for cooling and cleaning processes and in agriculture for irrigation. Agriculture is the biggest. Irrigation accounts for 70 percent of all fresh water used around the world and for more than 80 percent of water used in the United States. Poorly managed irrigation can cause water-logging and stalinization of the soil, while runoff of chemical fertilizers and pesticides contributes to water pollution.

Domestic use and the impact on the quality of people's lives are great. The Population Institute reported in 1990 that 1.7 billion people, nearly a third of the world's population, did not have an adequate supply of clean drinking water. Three billion people, more than half the total population, lacked adequate sanitation, flush toilets, sewers, sewage treatment plants, or alternatives such as efficient and safe waste collection. In many places, raw sewage was and is discharged directly into supplies of drinking or irrigation water.

Health and water are closely related according to the United Nations World Health Organization. Approximately 80 percent of all human disease is related to unsafe drinking water and poor sanitation.

The supply of fresh water is not endless. There is a limited amount of fresh water in the world, and water that has been contaminated with pollutants is no longer suitable for most uses. Water that drains into the ocean cannot be used for farming or drinking until it has evaporated, condensed into water vapor, and fallen as rainfall. Unfortunately, the places that receive the most rainfall are not the places that most urgently need it. Surface water sources, such as rivers and lakes are replenished steadily if rainfall is regular, but groundwater sources, or aquifers, are replenished so slowly that they must be considered resources, like oil and coal.

Walking In Dark to Get Water to Drink During Day

California, Nevada, Arizona, New Mexico, Florida, and Texas have experienced water shortages, but in many countries people cannot get drinking water in their homes. Some must line up at public taps. In Madras, India's "fourth largest city," the taps flow only between 4 and 6 a.m. and people must line up during the night in order to get water. As many as 8,000 rural villages in India have no water supplies at all. Each day the village women must walk for miles to the nearest well or stream. Many African women also carry their water jugs across miles of countryside every day to get water. In Mexico City, which has 22 million people and very little rainfall, the water level in the city's main aquifer drops 11 feet each year.

If it is not overused, the rate of recharge and discharge balances out, as do other phases of the hydrologic water cycle. But in most countries, the rate has not been balanced. Heavy withdrawals were concentrated in a few areas. Nine states accounted for 72 percent of the Nation's total groundwater use. California withdrew almost 15 billion gallons each day, Idaho and Texas used 7.4 to 7.6 billion gallons per day, followed by Arizona, Arkansas, Colorado, Florida, Kansas, and Nebraska. Irrigation accounted for 64 percent of all water used and public supply 19 percent.

Different Rates of Contamination

All pollutants do not result in the same rate of contamination for the same amount of pollutant. Groundwater is affected by the following factors:

1.   The distance between the land surface where pollution occurs and the depth from the water table. The greater the distance, the greater the chance that the pollutant will biodegrade, be absorbed by, or react with soil minerals.

53

2. The mineral composition of the soil and rocks in the unsaturated zone. Heavy soil and organic materials decrease the potential for contamination.
3. The presence or absence of biodegrading microbes in the soil.
4. The amount of rainfall. Less rainfall results in less water entering the saturated zone causing lower quantities of contaminants.
5. The evaporation rate. This is the rate at which water is discharged into the atmosphere as a result of evaporation from the soil, surface water, and plants. High rates reduce the amount of contaminated water reaching the saturated zone.

In the EPA's National Water Quality Inventory in a 1998 report to Congress, 31 of the 37 reporting states identified the types of contaminants they found in ground water. The states said that nitrates, metals, volatile, semi-volatile organic compounds, and pesticides were pollutants found most commonly in the water.

Factories Affecting Groundwater Contamination

Arsenic is a naturally occurring element in rocks and soils, and is the twentieth most common element in earth's crust. In groundwater, arsenic is largely the result of minerals dissolving from naturally weathered rocks and soils over time. The nation's groundwater typically contains less than one or two parts per billion of arsenic. Parts per billion are equal to about one drop in an Olympic-size swimming pool. Moderate to high arsenic levels do occur in some areas throughout the nation in a pattern related to geology, geochemistry, and climate. Elevated arsenic concentrations in ground water are commonly found in the West, parts of the Midwest, and the Northeast.

Arsenic research has shown that humans need arsenic as a trace element in their diet to survive although too much arsenic can be harmful. Arsenic can contribute to skin, bladder, and other cancers after prolonged exposure. The Environmental Protection Agency proposed lowering the current maximum contaminant level for arsenic in drinking water of 50 ppb to 5 ppb by 2006. Some scientists and public health officials believe that the level is too low and the treatment necessary to meet this standard will place an unnecessary financial burden on water suppliers.

Many scientist and geologists consider nitrates to be the most widespread ground water contaminant. Many states use its presence as an indicator of human impact on groundwater quality. Generally, a level of 3 ppb or more is considered indicative of human's impact.

Hazardous Waste

Every year, "270 million tons," of waste classified as hazardous are produced in the United States. That is more than a ton of hazardous waste for every man, woman, and child in this country. Ninety percent of all hazardous waste in the United States is produced by about 14,000 large waste generators, facilities each producing over 1,000 kilograms of hazardous waste a month. The chemical industry is by far the largest producer, followed by petroleum refiners, then by the metal processing industry. The remaining 10 percent comes from more than 1,000,000 small-quantity generators, businesses that each produce less than 1,000 kg of hazardous waste per month, photo labs, service stations, dry cleaners, body shops, printers, laboratories, and private homes.

The known sources of contaminants in ground water include disposal of hazardous wastes in landfills and industrial surface impoundment's, pits, settling ponds and lagoons, septic tanks, cesspools, and the chemicals used to clean them, municipal wastewater; mining activity, petroleum exploration and development, underground injection of wastes, agricultural, and street and urban runoff.

Water is a necessary resource for sustaining life. An adult human body is about 65 percent water and blood is 90 percent water.

Not only is water very important, but the quality of water is also very important. Ninety seven percent of urban homes get their water from public or private water systems, rural homes 4 out of 5 get their water from individual wells, and three-fourths of American homes are connected to public sewer systems. Twenty-four percent use septic tanks or cesspools.

Affecting surface and groundwater, pollution has severely limited the amount of drinking water available. In New Jersey leakage of chromium, copper, zinc, and other toxic metals and chemicals from industrial lagoons have contaminated millions of gallons of groundwater. New Jersey's Department of Environmental Protection made a study conducted for the National Cancer Institute and surveyed public wells serving more than 1,000 customers.

The results of these tests are sobering, said Thomas Burke, director of the study. The majority of the finished drinking waters contained low levels of potential cancer-causing volatile organics. The results clearly indicated the sensitivity of our water resources to the threat of chemical contamination. Evidence of toxic contamination was found in urban, industrialized areas as well as the most rural parts of the state.

The Environmental Working Group found herbicides in drinking water of 29 cities in the corn belt. Levels in a single glass of water contained nine kinds of herbicides in Fort Wayne, Indiana. In Danville, Illinois, the level of the weed-killer cyanazine was 34 times the federal standard, and in 18 sites, levels exceeded federal standards.

The United States Geological Survey estimates that "six billion tons" of hazardous wastes have already been dumped on United States land nationwide, and "40 million" additional tons are being added every year. There are about 7,000 hazardous-waste dump sites, another 200,000 chemical disposal sites, 200,000 municipal landfills and countless septic tanks, chemical spills, and other threats to clean water across the country. More than 700 organic chemicals used in industrial processes have been detected in our underground water supplies, including 22 known cancer-causing agents, dozens of other toxic substances, and many more that have not yet been tested for their effect on humans.

Toxic chemicals at thousands of hazardous waste sites across the country continue to seep into the nation's groundwater, contaminate the land, and poison the air. As of December 31, 1984, EPA had identified 19,368 hazardous waste sites of which 538 had been designated as priority sites. The EPA estimated that as many as 1.3 million sites might be "discovered" and evaluated to determine if they are problem sites according to the General Accounting Office, March 26, 1985.

Hot Water from Plant Killing Fish

In a 13 square-mile area of water that straddles Rhode Island and Massachusetts, 15 species of fish have all but disappeared in the span of a decade. The drop has been so complete that some fishermen call it, the dead zone.

Brayton Point, which is the largest fossil-fuel power plant, is sitting on the edge of the bay. The plant uses nearly 1 billion gallons of water daily and pumps much of it into the bay at temperatures up to 95 degrees Fahrenheit causing temperature increases of up to five degrees in parts of the bay.

The plants owner, PG & E National Energy Group, denies its cooling system has any connection with the disappearance of the bay's fish, and has spent at least $4 million since 1998 for scientific studies to prove it is not responsible. But the United States Environmental Protection Agency says it considers the plant partly responsible and has spent four years getting enough data to make an airtight case.

The Environmental Protection Agency was two years late in issuing a water-discharge permit to the plant. Brayton Point now supplies roughly one-fifth of Massachusetts' electrical power, and to cool its equipment, it uses water from the Taunton and Lee rivers that flow into Mt. Hope Bay.

The Rhode Island Department of Environmental Management said the plant increased it water use by 45 percent in 1985 and immediately fish stocks underwent a big downward slide. Stephen Medeiros, president of the Rhode Island Saltwater Anglers Association, said no one fish there anymore. The bay's study said the overall fish population dropped 86 percent after the plant increased its water use.

The huge intake pipes kills millions of fish larvae and eggs. And by raising temperatures in the bay, the hot water pumped out by the plant makes it inhospitable to some species. The company hired a fishery professor as a consultant, and gave a $1 million unrestricted grant to the University of Massachusetts at Dartmouth last year to document why fish populations are declining and to recommend solutions.

Not all studies are in, but those scientists say the plant does not appear to be having a dramatic effect on Mt. Hope Bay. PG & E also downplayed the importance of its intake pipes on the bay's fish population, arguing that few eggs and larva make it to adulthood anyway. The EPA announced a proposal to let power plants nationwide replace fish sucked into intake pipes, rather than find ways to keep them out.

Water for Hydroelectric Power

In 1995, 3,160 billion gallons of water per day were used for hydroelectric power generation, where falling water drives the plant's turbine generators and they generate electricity. All water withdrawals come from surface sources, which include reservoirs. No water is consumed directly during the cooling process, but return flow, water that is supplied but not consumed, may be less than 100 percent due to indirect factors such as evaporation from reservoirs that supply the power plant.

The Pacific Northwest used the most water for hydroelectric power in 1995. Three states with many natural waterfalls, Oregon, Washington, and New York, used about 46 percent of the water to generate hydroelectric power in the United States. In flatter terrain in the United States, dams create artificial waterfalls, allowing hydroelectric power generation. Building dams for hydroelectric power generation has had severe environmental impacts on river and stream systems.

We Waste To Much Water

A study by a Detroit Free Press found that "waste and artificially low prices for water are the real problems in Arizona." It is not just the swimming pools, man made lakes, un-metered sprinklers for lawns particularly in Phoenix, and enormous fountains, including the biggest one in the world, which shoots water upward at 7,000 gallons a minute at the Fountain Hills desert development. Agriculture, which uses 90 percent of consumed water and about three fourths of it is in Central Arizona for crops whose production the government is limiting because of "over production." Many Arizona farmers use the water on inferior land to grow such crops as sorghum and alfalfa that require large amounts of water. The rest of us are paying for it.

Senator Dave Durenberger R-Minnesota says, "It makes no sense to spend billions of Federal dollars to irrigate semi-arid lands and then spend billions more to buy the crops because there are no markets." The Washington Monthly reported that an array of tax breaks and farm subsidies underwrites plowing and irrigating lands ill suited for crops. Federal policy encourages enormous waste by providing water for irrigation at prices that cover as little as 2 percent of costs.

Both the quality and quantity of water resources need to be protected for the nation's present and future generations. Although present water use can be determined, total water needs for most uses are changing. Water use is dependent on prices, technology customs, and regulations. Even though water-use data are good indicators of where and how the nation consumes water today, they are not necessarily good predictors of future water-use trends.

For much of the country, the era of free and easily developed water supplies has ended. In some areas, water use is approaching or has exceeded the available supply. In most areas, the nation is running out of water. Hydrologists, scientists who study the properties, distribution, and circulation of water, engineers and economists must provide a wider knowledge of how to use water more efficiently to fit regional demands. Water users must treat water as a valuable natural resource that cannot be carelessly wasted.

A large share of the Federal irrigation program's benefits goes to corporations running farms as large as 20,000 acres. In the San Joaquin Valley of California, water is provided for land owned by Getty Oil, Tenneco West, and JAG. Bowel is a huge cotton corporation. They are the companies that pay a large amount of money lobbying to get what they want, but pay little for the water.

Water from Federal projects costs are so small, comments the Washington Monthly, that farmers find it cheaper to use than to save water. In DA's Wetlands reclamation district, where the average farm is 2,400 acres and produces profits of half a million dollars a year, the Federal government is charging $10 per acre foot. In neighboring areas, water on the free market may cost 100 times that amount or $1,000. In South Dakota, users pay $3.10 an acre-foot for water that actually costs $131.50 to produce. If this seems impossible, think about what effect a $500,000 check would mean to a politician.

People in the United States and other industrialized countries must learn to waste less water. They can install toilets and shower heads designed to save water, or plant their lawns with vegetation adapted to the local climate so that watering will not be necessary. If everyone on earth started using water as efficiently as possible, the demand may still out grow the supply if it has not already done so.

In 1975, 19 countries in the developing world did not have enough renewable water resources for basic domestic uses and irrigation. By 2000 that number was expected to reach 29, and by 2015, 37 nations will be experiencing severe water stress. In the late 1980s, the World Health Organization claimed that 80 countries, home to 40 percent of the world's population, were already experiencing some degree of water shortage. With global population accelerating toward 12 to 14 billion in the 21st century, maintaining dwindling water supplies and reducing contamination must be viewed as strategic priorities.

Because ground water lacks the self-cleaning properties provided to surface water by dilution, circulation, and degradation by sunlight and aquatic organisms, ground water can remain contaminated for centuries.

How to Control Water Use in Our Homes

The first step is to repair or replace aging leaky water supply systems and outdated treatment plans in most industrial cities. Some of which pump 40 percent or even 50 percent more water than they can account for in billing. New York City lost at least 100 million gallons of the 1.5 billion gallons it pumped every day through 6,000 miles of piping, half of which had been in place for 45 years or more. Three major water mains in the city have burst since August 1983, and the related costs, to say nothing of the wasted water, were staggering.

An environmental group reviewed 19 cities and discovered that many of the treatment plants were using nearly century-old technology and were not up to the task of cleaning up contaminants, said Erik Olson who is the author of the report by the Natural Resources Defense Council. Pipes carrying water often are old and in some cities dating back more than a century.

Chicago had a rating of excellent for its tap water quality and New Orleans, Denver, Baltimore, Manchester, N.H. and Detroit were rated good.   Five were given poor ratings, Phoenix, Boston, Albuquerque, Fresno California, and San Francisco.   Some people would be surprised to know that their water may contain cancer-causing chemicals.   Toxic chemicals like lead that often contains the remnants of pollutants like sewage that slips through some of the treatment plants Olson said.   The group says arsenic, rocket fuel perchlorate which is a particularly dangerous chemical, and chromium can enter the water from industry and manufacturing.   The rocket fuel pollutant was found in the Colorado River that serves a lot of cities.   The estimate is $500 billion over a 20 year period to fix the systems.   Again America is looking at what it will cost to make an attempt to correct the problem rather than how many lives might be saved.

Up to 45 billion gallons of water per day could be saved by adopting innovative irrigation techniques, such as that developed in Israel which uses "drip irrigation" to produce almost twice the land's former yield of melons while using less water to do so.   Water is a life or death issue.

Martindale and Peter H. Gleick described a massive water conservation project undertaken in New York.   To prevent a pending water crisis in the early 1990s, New York city needed and extra 900 million gallons of water per day, about 7 percent of the city's total daily use.   Faced with the need to raise $1 billion for a new pump station to bring additional water form the Hudson River, the city came up with a cheaper alternative by reducing the demand of the current supply.   Using a three year toilet rebate program, budgeted at $295 million up to $1.5 billion rebates, the city sought to replace about one-third of the existing toilets that used five gallons per flush with the water-saving models that did the same job with less than two gallons.   By the end of the program in 1997, 1.33 million inefficient toilets in 110,000 buildings had been replaced with low-flow toilets.   The result was about a 29 percent reduction in water use per building per year.   The low-flow toilets were estimated to save about 70 million to 90 million gallons per day.

Also inspectors checked for leaky plumbing, offered advice on retrofitting with water efficient fixtures, and distributed low-flow showerheads and water-efficient faucet aerators. Low-flow shower heads used about one-half the amount of water the old units used. Faucet aerators, which replaced the screen in the faucet head and added air to the spray, reduced the flow from four gallons per minute to one gallon per minute. The company made several hundred thousands of these inspections, saving an estimated 11 million gallons of water. Per person water use in New York City dropped from 195 to 169 gallons per person per day between 1991 and 1999, but the city keeps growing.

Different Ways We Use Water

Water plays a very important part in everyone's life, but we pollute it every day in many ways. At one time Americans were not concern about the results of pouring wastes and emptying garbage into rivers, lakes, or the ocean. Water was so vast they thought it could easily absorb all the garbage that was put into it and no one really cared.

As America grew bigger and bigger and cities increased in size, pollution worsened. Factories poured their wastes directly into the rivers. Big cities with huge amounts of garbage dumped it into big holes in the ground where pollution formed. The residue from the garbage seeped below the surface into ground water and many cities poured their sewage directly into the rivers. Other's loaded their garbage onto big barges that sailed out into the ocean and dumped their garbage.

Farmers wanting to raise more food for the growing nation returning greater profits, started using increasing amounts of pesticides and fertilizers. When it rained some of the pesticides and fertilizers ran off with the extra water into the lakes and rivers. The streets became coated with oil and other chemicals from cars and trucks, and homeowners used fertilizers to make their lawns grow and when it rained oil and fertilizer ran off with the rain into storm sewers. Melting snow also carried pollutants, including the salts used to melt ice on streets and roads into storm sewers that emptied into rivers or lakes.

Many parts of the coast show the result of the abuse of the environment. Some previous studies have reported more than ten tones of garbage per mile of coastline. Many things are thrown into the water and over half of the garbage is made of plastic. Much consists of metal, glass, plastic foam, paper, bags, lids, bottles and rings used to hold six packs of beer and soda together. Also people discarded toys, egg carton, shoes, diapers, and even hard hats in the ocean waters.

Most of the garbage does not float and sinks to the bottom of the ocean where it stays, fish try to eat it and other sea life or is eventually disintegrated and further pollute the oceans.

Floating ocean garbage comes from manufacturing plants that pour their wastes into the ocean, city sewer systems dump their sewage, garbage barges throw away the trash from the cities, freighters toss trash over the side as do many people living and working on offshore oil and gas rigs. Commercial fishermen leave old nets behind or some of their fishing equipment falls apart and disappears into the ocean. Swimmers toss their trash and people in boats throw their trash overboard.

Wetlands

The term "Wetland" encompasses a variety of wet environments, coastal and inland marshes, wet meadows, mudflats, ponds, bogs, bottomland, hardwood forests, and wooded swamps. It is always or often saturated by enough surface or ground water to sustain vegetation that is adapted to saturated soil conditions, cattails, bulrushes, red maples, wild rice, blackberries, cranberries, and peat moss. They are rich in minerals and nutrients and provide many of the advantages of both land and water environments, and have a diversity of species, including many insects that are a basic link in the food chain.

Wetlands have been regarded as foreboding and dangerous places which had little economic value. They were pushed to be destroyed or developed and more than 100 million acres of the nation's wetlands have been destroyed. From the mid 1950s to the mid 1970s such losses averaged 458,000 acres a year. That means that an estimated 54 percent of the wetlands that existed in colonial times have vanished forever.

The hardest hit areas were inland marshes and swamps, the vegetated wetlands considered most valuable. During the two decades covered by the study, 6 million acres of forest wetlands, 400,000 acres of shrub swamps, 4.7 million acres of inland marshes, and 400,000 acres of coastal marshes and mangrove swamps were destroyed. More than 11 million acres had disappeared, a total area twice the size of New Jersey.

More recently, we have come to realize that wetlands are precious ecological resources. Resources that nurture wildlife, purify polluted waters, check the destructive power of floods and storms, and provide all kinds of recreational activities. This new attitude is reflected by two decades of Federal and State laws and other programs that serve to preserve and protect our remaining wetlands. They reduce flood crests and flow rates after rainstorms.

Sediment is deposited in stream beds, wetlands, and bays as a result of dam and levee construction, terracing, the clearing out of drainage ditches, and to a lesser extent from general cultivation.

These areas are ecologically critical. They serve as spawning and nursery areas for commercial and sport fish, act as natural cleaners for airborne and waterborne pollutant and supply essential nesting and wintering areas for waterfowl, and sediment deposits, especially those containing pesticides, may have adverse impacts on them. As yet, little information is available on the nature and extent of these deposits and their harm.

In the 1970s there was a total of 99 million acres of wetlands or about 5 percent of the nation's land surface left in the United States and 93.7 million acres consisted of inland freshwater marshes, swamps, bogs, and ponds. The remaining 5.2 million acres were made up of coastal saltwater marshes.

The rapidly expanding demands of agriculture accounted for 87 percent of this great loss. Eleven point seven millions acres of wetlands had been drained for crop production. At the same time agricultural development and construction projects created 2.1 million acres of ponds and 1.4 million acres of lakes, but not nearly as productive or valuable as the vanished vegetated wetlands.

One of the important aspects not measured by the study was the deterioration of many wetlands. The reduced quality of wetlands stems from many causes, including pollution from rivers, streams, adjacent fields, urban encroachment, building of highways and railroad roadbeds, the construction of ditches for mosquito control, and oil and gas development canals that allowed saltwater intrusion into freshwater marshes.

Our most threatened wetlands were inland marshes. The many thousands of shallow water filled depressions in the upper mid east states and central Canada are called "prairie potholes." These gouge marks, left by the retreating of the last ice age, serve as a vast waterfowl breeding ground. The prairie regions of the Dakotas, Northern Montana, Nebraska, Iowa, Minnesota, and Southern Canada are a principal nesting area for North America's migratory waterfowl.

A net loss of vegetated wetlands, inland marshes, forested wetlands, and shrub swamps of 11 million acres, are nearly all due to agriculture. An overall net gain in non-vegetated wetlands, ponds and inland mudflats, of 2.3 million acres, are due to the building of farm ponds. A net gain in deep water habitat, inland lakes, reservoirs, and 1.4 million acres is a result from construction projects.

Trends in Wetlands

A net loss in vegetated wetlands, coastal marshes of 372,000 acres are mostly due to conversion into the open water of bays and sounds. An overall net gain in sub-tidal deep water, bays, and sounds, of 200,000 acres.

The prolific wetlands are scattered throughout some of the nation's most fertile agricultural areas. Consequently, a great amount has been drained and turned into crop land. Nebraska's Rainwater Basin is a focal point in the central flyway that is used by millions of ducks, geese and cranes during their annual migrations, but 90 percent of the original wetlands in the basin have been destroyed.

The inland marshes of Florida provide both feeding areas for wading birds and wintering grounds for waterfowl. In addition, they supply breeding habitat for such species as rails, the mottled duck, and the endangered everglade kite. These wetlands are also prime habitat for furbearers, alligators, and various other kinds of wildlife. The conversion of the marshes to agriculture is significantly affecting both water fowl and other wildlife populations.

Nearly 80 percent of the 25 million acres of periodically flooded bottom land hardwood forests that once existed in the Lower Mississippi Valley has been lost to agriculture. The remaining areas still serves as the major stop over wintering grounds for most of the continent's mallards and for virtually all of the wood ducks in the central United States. They also provide rich habitat for a wide range of other wildlife and spawning, and nursery areas for fish. By 1977 only 5.2 million acres of bottom land hardwood forests were left in the Mississippi Delta. Even today these remnants are shrinking as tracts are leveled, drained, and converted for soybean farming.

Wetlands of coastal North Carolina are covered with evergreen trees and shrubs. They serve as a critical function by regulating the flow of fresh water to nearby coastal estuaries. This flow of fresh water is essential in maintaining the Palmetto Sounds productive fisheries. Palmetto is under great pressure from peat mining for fuel and the other wetland for agriculture.

The causes of the destruction have been both natural and man made. Most of Louisiana's losses are due to sinking terrain and subsequent flooding. These conditions stem from a variety of factors such as a rise in the sea level, subsidence of the coastal plain, levee construction, canalization, and oil and gas extraction. Directly contributed to man has been due to urbanization, daily dredging, and developments. As the space needed for cropland continues to grow and urban areas continue to expand, America's wetlands will continue to shrink.

Therefore, it is more important than ever to monitor and propose alterations of wetlands. Man always has to satisfy his profitable wants.

Chapter 4: Industries Pollution

Industries Operation Causing Cancer

The "Love Canal" in Niagara Falls, New York, is known for its toxic contamination and cleanup of water that sustained life. In an inactive canal, the Hooker Chemical Corporation and Olin Corporation disposed 20,000 tons of toxic chemicals. When the canal was full in 1993, the wastes were covered by layers of soil and clay and the land was given to the city where a school and playground were built. Apparently their theory was, if water does not penetrate the chemicals, there will be no leaching.

Early, water integrated with the deadly chemicals and began to kill vegetation and seep into basements in 1976. There were birth defects, miscarriages, epilepsy, blood diseases, and cancer. Jimmy Carter was president and declared the Love Canal a disaster area. One thousand families were evacuated.

Jacobson Shipyard in Oyster Bay, New York in 1991 was discovered to contain concentrations of lead that was 36 times greater than legal levels. During World War II no one looked into the future and the possibility of contamination even though the company sandblasted tugboats and ships. It would have been easy to think about the copper paint and the chips of lead that would fall into the water. Since America is always putting a cost on clean up and overlooking the results of the contamination, it cost $6 million and took five years to clean up the area after damages to health had been done. The Amazon River in Brazil is being poisoned by mercury used by gold miners verified by fish samplings and hair samples of miners that show toxic levels of mercury.

The cost of dredging a lake bottom in Chicago where lead leached into a lakebed from a neighboring gun club is expected to be $1 million dollars. Will it be removed or will a large portion of it be relocated?

The Monsanto Company in Gadsden, Alabama polluted the town of Anniston with PCBs in 2002. The company had a $40 million settlement in a federal case last year and has also spent tens of millions of dollars trying to clean up contamination in and around the town. The company also paid $43.7 million to 5,000 Anniston-area property owners along Choccolocco Creek and Lake Logan Martin where PCBs were found. It was the first from a jury over the PCB contamination in Anniston, but the situation had dogged Monsanto for years.

Monsanto produced PCBs for nearly a decade up to early 1970 when the chemical was banned as a suspected cancer-causing agent. Its major use was insulating fluid in electrical transformers. David Baker said the St. Louis-based company caused the death of his brother two decades in the past, but his case had not been heard. He said as a child growing up in that area, the odor was so bad from the plant that we had to go inside the house. People could not work in their garden and children could not play in their yards.

A Montgomery Law firm has about 15,000 clients suing in federal court, and 3,500 Anniston residents and business owners sued the companies claiming Monsanto knowingly contaminated their community. Most of the homes and businesses around Anniston are vacant and other properties have a negative value said Gene Tomlin that operates Anniston Quality Meats near the chemical plant.

A chemical factory in Minamata, Japan dumped mercury into the sea between 1953 and 1968. Many died and suffered sever neurological damage from eating shellfish contaminated with methyl mercury. Toxic metals and pesticides in seafood pose a possible more serious health threat than viruses and bacteria because their effects can be permanent. Health officials recommend limiting intake of a variety of fish and shellfish due to high concentration of metals and pesticides.

There were 2,262 lead poisoned victims severely and permanently handicapped and were eventually paid $60,000 with medical expenses paid. Milder symptoms were recognized in 1996 for 10,000 people and they were given $24,000 each by the company.

In a Missouri river town named Herculean, pollution kept children inside. On a fine spring day, sunlight floods the cozy bungalows along the rolling bluffs of the Mississippi River town. The wind whistles through the vacant Little League ballpark. Backyard swings gently click in the breeze, and dog leashes lie curled on the ground. The town looks deserted, but it is not.

Signs caution parents to keep children off sidewalks, out of driveways, and out of streets. They advise residents to wash their hands when coming inside before eating, and again at bed time. The 2,800 residents in the community, a half-hour south of St. Louis, have long suspected that they might be at risk. But only in the past few months have they begun to realize the extent to which their river town is contaminated with dangerous levels of lead, arsenic, cadmium, zinc, and sulfur. Mountains of lead slag piles stand in a wetland and leech contaminates into a river where people used to fish.

The source of contamination is the town's 110-year lead smelter, the largest in the nation, operated by Doe Run Company. Blood tests conducted by Doe Run in the past couple of months indicate that more than 55 percent of children age 6 and younger who live with in a half-mile of the smelter had dangerously high lead levels and the 30 percent of children under 6 with in one mile of the smelter also had high lead levels.

Doe Run has evacuated a few residents, but the majority remain at home, inside, and wait. All the residents can do is follow tips such as to remove shoes once inside the home. Doe Run gave some residents high-powered vacuum cleaners to suck up contaminants. Others had their homes sanitized by contractors hired by the lead smelt company. Many have had their yards stripped one foot deep of soil and the dirt replaced. A few have had their yard dug up twice.

Doe Run bought out some 60 homes near the lead facility and then rented homes to people like Sadowski, who moved in nine years ago when her children were ages 3, 4, 5, and 9. She cannot tell whether her son's inability to concentrate is from contamination or just adolescence.

Lead poisoning can cause learning disabilities, behavioral problems at very high levels, seizures, coma, and even death, according to the federal Centers for Disease Control. Doe Run agreed to buy 26 homes immediately, and is expected to sign an agreement to buy 160 homes, possibly over a period of two years. The company will not pay to relocate Sadowski's family, but did provide them with a high-powered vacuum cleaner. She was told that her yard will be stripped again, for the second time in five years. They keep a 3 year old entertained in the house but she want to play outside. I explain, but she does not understand. I cannot tell you how many times she's beat me at Cancy Land.

Madison's lead level was 19 micrograms per deciliter. The federal standard for an elevated lead level was 10. Anything higher is considered dangerous to the nervous system, brain, and organ development. Most of them could not afford to move because of the expense.

Blood testing had focused on children under 6. Younger children are more likely to take in the lead because of increased hand-to-mouth activity, according to the CDC. Women in their child-bearing years and particularly pregnant women are at high risk. Not only does lead damage the woman, but also the fetus.

The Environmental Protection Agency in 1993 said, 819 systems serving 30 million people had excessive lead levels in their drinking water. It was surprising to know the source of the lead was home plumbing systems. Most of the plumbing used now is copper.

Army Burning 31,500 Tons of Nerve and Blister Agents

The Army is scheduled to destroy about 31,500 tons of nerve agents and highly toxic blister agents the last of 2002 at a projected cost of $24 billion. Tooele, Utah and Johnston Atoll in the Pacific Ocean destroyed about one fourth of the stockpiles. A report said it would be done despite chemical releases and violations at the only two operational incinerators.

The National Research Council, a branch of the National Academies of Science said, the risk to the public and to the environment of continued storage overwhelms the potential risk of processing and destruction of stockpiled chemical agents. The destruction of aging chemical munitions would proceed as quickly as possible.

About 40 cases of chemical agents leaked into areas where it was not supposed to and it escaped from an incinerator building. A council's report said amounts that escaped were too small to threaten the public. How does a group decide what amount is too small and what amount will cause a fatality? The major hazard to the surrounding communities is the potential releases of agent from stockpile storage areas they said.

The report said many of the munitions and rockets that hold the chemical agents are aging and leaking and detonation of the stored chemicals could spread a large amount of agent into the atmosphere.

Charles Kolb is chairman of the committee and made a statement that the technology is capable of doing the job if it is run correctly, and there are no reasons it cannot be run correctly if management puts its mind to it and trains its work force properly. After a statement like that, you wonder about the concern and the efficiency that will be applied.

Craig William is director of the Chemical Weapons Working Group based in Kentucky and favors chemical neutralization to incineration. He said the report ignored thousands of pages his group submitted to document incinerator problems and glossed over worries voiced by local officials and complaints by whistleblowers. Charles Kolb said his group considered the information submitted by William's group, but it was largely repetitive or undocumented.

A stockpile of the nerve agent sarin has been incinerated and the Army is preparing to burn the more toxic nerve agent VX. Incinerators in Anniston, Alabama, Pine Bluff, Oregon, and Umatilla, Oregon are scheduled to begin operations in the coming months. Chemical agents at Bluegrass, Kentucky, New Port, Indiana, Aberdeen, Maryland, and Pueblo, Colorado are to be neutralized using chemicals rather than incinerations.

The Cost of Dredging

No one is betting that industrial pollutants like PCB will disappear. Boston Harbor is only one of many coastal waterways whose bottoms may have been permanently contaminated with chemical waste and toxic metals. Other areas with especially difficult toxic waste problems include New York Harbor, parts of the Mississippi River estuary in Southern Louisiana, and parts of Galveston Bay. Three areas of Puget Sound, the vast inland sea that is the pride of the Pacific Northwest, are heavily polluted and have been designated toxic-hazard sites on the EPA's Supervened Cleanup List. The bottom sediments and local flats of Commencement Bay, and inlet bordered by oil refineries chemical points, pulp mills, and a defunct paper smelter, are loaded with petrochemicals, copper, lead, zinc, and arsenic. Nearby, Eagle Harbor is so polluted that much of its rapidly diminishing produce and the area's harbor seals, are registered as one of the highest levels of PCB contamination ever recorded.

Bay bottom contamination by heavy metals and organic chemicals poses an environmental double whammy. Many such pollutants become more concentrated as they travel up the food chain. The obvious conclusion is that fish from polluted waters should not be eaten. New York state officials asked consumers not to eat more than one serving per week of any fish taken in coastal waters, the Hudson Rivers, or Lake Erie. Women of childbearing age and children under fifteen are advised not to eat many New York state fish at all. We have been told that red meat is not good to eat and that we should eat fish and poultry.

The other problem with contaminated bottom sediment is that there is no good way to remove it. Some pollutants, such as DDT and PCB are extremely long-lasting chemical compounds that have spread very widely in rivers, bays, and tidal muck. Dredging the bottom only spreads them further, which is why environmentalists often oppose dredging projects for navigation or development though some officials now say polluted sediment can be capped with a layer of clean fill. That remedy is probably useful in only a few relatively small problem areas, if any.

Every year, millions of tons of materials are dredged from the bottom of harbors and coastal areas to clear or enlarge navigational channels or for development purposes. Dredged material usually contains high concentrations of pesticides, metals, and toxic chemicals. During the dredging process, or when the dredge material is dumped, pollutants that had settled into the sediment are stirred up and released into open water and have a greater potential to harm resident marine life.

Lakes are very susceptible to pollution because of their ability to capture pollutants and accumulate them within their water, bottom sediments, and aquatic life. It is less cost effective to clean up lakes and rivers after they get polluted than to prevent them from being degraded. The Longer Congress delays in addressing priority non-point source problems, the more expensive it will be for our children and grandchildren to restore suitable quality to our Nation's waters.

Flood damage and dredging costs are extremely high. Certainly several hundreds of millions of dollars have been expended to keep water open for navigation over the last five years. Federal water projects such as Fish Trap and Dewey Reservoirs in Kentucky are filling with sediment. In the West, sediment is deposited in

Federal water projects and displaces water that could be stored for irrigation. The state of Wyoming has identified Boysen Lake as such an example suffering from accelerated sedimentation. Urban runoff and sewer leakage have caused public health concerns about bacterial contamination across the country. Fish are contaminated in urban areas as well as rural.

Some pesticides, especially the chlorinated hydrocarbons, are very toxic and very durable. DDT, banned in the United States 20 years ago, but still used in other countries, is known to destroy egg shells, thus reducing bird populations. It is still found in ocean sediment and in tissues of many fish and marine mammals. Toxic pollutants eventually accumulate in the bodies of fish-eating land animals and up the food chain.

A Sewage Conduit and Sewage Tunnel

Close to the eastern shore of Deer Island in Boston Harbor, construction crews were working around the clock blasting and drilling a 9.5 mile sewage conduit. It would become the main drain for Boston and 42 suburban communities that send their sewage toward the harbor.

They bored a 430 foot-deep shaft straight down into the bedrock below the island. Once they reach that depth, they assembled a boring machine that is as long as a football field to drill a tunnel 9.5 miles out to sea under the ocean floor.

Five years from 2001 the tunnel will be the longest in the country that is not used for vehicles and will be nearly the diameter of the Callahan Tunnel linking Boston to Logan Airport. It has 55 vertical riser or diffuser pipes to spew the treated effluent into the ocean and will cost nearly $310 million. Eventually, it will carry 1.2 billion gallons daily, an amount equivalent to the total flow of the Charles, Mystic, and Neposet Rivers.

It is part of the court ordered Boston Harbor cleanup, a $6.1 billion project that include primary and secondary sewage treatment facilities to replace aging facilities on Nut and Deer islands that provide only limited treatment.

Those overburdened plants, which discharged effluent less than a mile from shore in shallow water, are a major reason why Boston has long had one of the nation's most polluted harbors.

A second sewage tunnel work began on another major element of the project. Another 9.5 mile tunnel is to transport sewage from communities south of Boston. It runs from the collection point at Nut Island to the Deer Island facility. Two hundred fifty to 300 feet below the sea floor, cracks in rocks means some water leakage. The tunnel will be drilled on a slightly uphill grade, enabling seeping seawater to flow downhill to the point where the tunnel connects with the vertical shaft. The liner in the tunnel will be 10 inches thick.

The self-propelled drill is a custom-built machine whose face will cut a swath 26 & 1/2 feet in diameter. The face is studded with more than 50 nickel alloy steel disks that roll across the rock, cracking and crumbling it into 3 inch sections that are scooped into a bucket installed behind the cutting surface. At the end of the tunnel, 9.5 miles out to sea where most of the boring machine useful life is over, it will be encased in concrete and abandoned.

Along the last 6,600 feet of the tunnel, 55 riser pipes will be bored down from the sea floor to connect to the tunnel. When the job is completed, it will force the effluent upward and out the pipes into the sea. The risers, which encase a 30-inch-wide smooth fiberglass pipe, will be drilled two years before the tunnel is finished and will be lined up relying on satellite navigation systems to line them up parallel to the tunnel's course.

Real Estate

Fall River, in the sizzling real estate market of 1988, there was a middle aged couple living on disability checks that could afford a vinyl-sided raised ranch house close to power lines and a four lane highway. Maria and Bernardino Gigueiredo put down $50,000 on the home. The ground beneath the house had been home of a century's worth of industrial companies using toxic chemicals. But the contamination had been cleaned up, the Gigueiredo's were told, and the land had been certified for housing by the city of Fall River.

More than three years later, a routine state audit revealed a plume of potentially dangerous solvents still in the ground. Driven from their home by toxic fumes, the Figueiredos spent three weeks in a hotel and hired a lawyer to sue the developers who sold the house.

The case exposes what critics say are gaps in the state's decade-old system for cleaning up contaminated waste sites, a system that has resulted in the redevelopment of 1,250 contaminated sites and been touted as a national model.

They cheated us from the beginning, "Maria Figueiredo said through an interpreter. No one would every buy this house, the way it is now. The land had housed a bleaching facility, a webbing mill, a fur factory, an electrical contractor, and a commercial laundry.

Before 1993, developers building on such sites had to apply to the Department of Environmental Protection and follow steps carefully prescribed by the agency. In 1993, a new state law gave developers and polluters the authority to decide the course of cleanups. DEP would serve as a backstop, auditing paper work and ensuring that any remaining contamination did not exceed allowable levels.

But the real estate boom brought greater pressure to build houses on marginal properties. And the DEP had been stretched thin, causing advocates to think they could check cleanups adequately, thereby exposing home owners like the Figeiredos to contamination.

They hired a former DEP official named Theodore Kaegael Jr. Kaegael that had worked for the DEP for 17 years, five of those years as chief of the emergency cleanup response unit, before starting his own consulting firm. The Figeiredos' received some money and moved to another house.

Oil Spills

Another item that produces pollution is gasoline and motor oil. One gallon of gasoline can pollute a million gallons of water, and contaminants often found in used motor oil add to its toxicity. Approximately 600 million gallons of used motor oil are generated each year in the United States and about 90 percent of the used oil generated by do it yourself oil changes is improperly disposed of. The equivalent to 200 million gallons, equal to 18 Exxon Valdez spills, dumped on the ground, poured down the sewers, or thrown in the trash where it will ultimately taint water supplies.

On June 22, 1969 the Cuyahoga River in Cleveland burst into flames, the result of oil and debris that had accumulated on the river's surface. Five-story-high flames destroyed two bridges and put the problem of environmental pollution into the public consciousness via the nightly news and other reports in newspapers. In 1972, Congress passed the Federal Water Pollution Control Act known as the Clean Water Act.

The sight of dead and dying otters and birds covered with black film created sympathy. The size of the spill itself is often not the determining factor in the amount of damage it causes for oil spills. Other factors include the amount and type of marine life in the area and weather conditions that would spread the oil.

When the Exxon Valdez ran into a reef in 1989, 11 million gallons of oil spilled into one of the richest and most ecologically sensitive areas in North America. An oil slick the size of Rhode Island threatened fish and wildlife. Otters, which cannot tolerate even a small amount of oil on their fur, died by the thousands. Despite efforts by trained environmentalists and local volunteers to save them, oil-soaked birds lined the shores, only to be eaten by larger predator birds, who then succumbed to dehydration and starvation because the ingested oil destroyed their metabolic systems.

One of the things that came out of the Oil Pollution Act of 1990 was a requirement that new tankers have double hulls. When equipped with two hulls, if the tanker's exterior hull is punctured, the interior hull holding the oil will still remain intact. The law requires older tankers to be fitted with double hulls by the year 2010. The Exxon Valdez was neither double-bottomed nor double-hulled.

When the Norwegian tanker named Mega Borg spilled 4 million gallons of crude oil into the ocean, a new method of dispersing the spill was tried by several Texas agencies. The Bioremediation was the new technology and used billions of living microbes to consume the oil.

There are three types of bioremediation, nutrient enrichment, seeding, and the use of genetically engineered organisms. When the tiny organisms begin eating, they multiply until their oil food source, is depleted. The microbes then die and are eaten by other scavenger microbes. The resulting waste is a fatty acid that harmlessly re-enter the normal food chain. One hundred and ten pounds of microbes were sprayed over the 40 acres of oil in which there were over a trillion microbes in each gram. Within "hours" the dark brown areas turned mooted, then yellow, and finally disappeared.

Reduced Funds for Five Toxic Waste Cleanups in New England

The EPA was to cut millions of dollars out of superfund's. Five toxic waste cleanup projects in New England are slated to be scaled back or put on hold because the Bush administration has decided to slash their funding.

The EPA inspector general indicates that in the next fiscal year, two Superfund sites in Massachusetts and one each in Vermont,

New Hampshire, and Maine will receive no money or substantially less than they requested.

The New England sites are the most severely contaminated in the nation, but are facing a total funding shortfall of more than $47 million, part of a $227 million budget gap that affects 33 Superfund sites in 19 states.

The area around Fairhaven's Atlas Tack Corp plant in Massachusetts which manufactured wire tacks and steel nails in the 1970's, is contaminated with cyanide and arsenic, among other chemicals. EPA analyses have found that the plants toxic waste is threatening residents and wildlife as it seeps into a saltwater tidal marsh in Buzzards Bay. About $13 million was requested for the cleanup which is expected to cost $18 million and take two years. But the project is unlikely to receive any funding according to the inspector general's report.

A cleanup site in nearby New Bedford faces a multimillion-dollar funding shortfall. For decades, manufacturers dumped polychlorinated biphenyls, or PCBs into the Acushnet River, which empties into New Bedford Harbor. The project has received $6.5 million of the $10 million officials said they needed this year.

The New Bedford and the Atlas Tack are in the midst of economical depressed communities. They have been unable to attract new business because of the contamination, said Pam DiBona, vice president of the Environmental League of Massachusetts. Economic revitalization efforts will come to a halt if the Superfund cleanup is delayed much more.

Bush does not consider environmental protection a priority. There is no reason for the cuts since the EPA opened their checkbook for tax give away for the oil and gas industry said Senator John F. Kerry, Democrat of Massachusetts. "Taxpayers should not be left to foot the bill and suffer the public health consequences of dangerous toxins left behind by polluters."

Andrew D. Anderson

EPA officials had requested $8.6 million in 2002 to start cleaning up the New Hampshire Plating site in Merrimack, N.H., where toxic sludge has been detected near drinking water wells that serve about 39,000 people. No money was allocated.

At the Eastland Woolen Mill in Corinna, Maine, wool-dying chemicals have oozed into the Sebasticook River, which has been significantly contaminated according to an EPA analysis. This year, the cleanup project at the mill has received $5 million of the $12 million that regional EPA official said they needed.

As part of the clean up of contaminants from the factory, the entire down town has been razed and the river shifted to make room for a new main street. It was unclear exactly how the cuts would affect that project.

In New Ellenton, South Carolina in the Cold War in the 50s, people welcomed the opening of the bomb plant along the Savannah River. South Carolinians for decades have embraced just about any industry that could bring jobs to the countryside. But now that attitude is changing, as South, Carolina is required to clean up.

Burning Coal

Control before Combustion. The amount of sulfur and nitric oxides given off by coal can be reduced by applications of physical and chemical processes before combustion. Coal is normally cleaned at the mine to remove particles of sand, clay, and other impurities. Coal washing is a relatively simple process in which the coal is crushed and the particles put through large tanks of water. In this process 50-90 percent of the pyrite is removed from the coal, but removing the organic sulfur involves a much more complex and expensive chemical process, based on the use of microwave energy and electron beams. Chemical treatment can remove both pyretic and organic sulfurs, but none of the processes has reached the commercial stage of development.

A process that shows promise for the future in the removal of pyretic sulfur from crushed coal is the electrostatic process. When coal is fed into a rotating device from an electrostatic charge, the coal adheres to the drum, and the pyrites and ash are removed. Experiments indicate that this process, when perfected, will remove from 38 too 68 percent of the sulfur and from 50 to 60 percent of the ash.

Coal cleaning is now widely practiced throughout the world, and about 40 percent of the coal produced in the United States is cleaned. As a result it is estimated that sulfur dioxide emissions are reduced annually by 2.4 million tons. If all the coal was cleaned, sulfur emissions in the United States would be reduced by more than 5 million tons.

There is another aspect about coal as it is removed from the mine that may be interesting to the reader. Machines with big teeth dig coal in underground mines and funnel it automatically on to conveyor belts. It was initially dumped outside the mine where it was loaded onto trucks to be hauled to processing plants, rail yards or river ports.

In Louisiana, American Electric Power moves lignite coal 4 miles to a power plant near Shreveport completely by conveyors. No trucks are used. Conveyor belts carry coal up to 9 miles through underground mine shafts and tunnels where it is made ready for shipment to electric-generating plants where it is usually washed before being used.

Jobs vs. Contamination

The United States Department of Energy announced plans last fall to ship weapons-grade plutonium from its Rocky Flats installation in Colorado to the Savannah River Site, where it would be converted into nuclear reactor fuel over the next two decades in an operation that could create up to 800 jobs. But Governor Jim Hodges has said he does not trust the government to keep its word and fears the plutonium will be left at Savannah River permanently.

He vowed to do "whatever it takes" to prevent the radioactive material from being stored there. The Highway Patrol has conducted drills on how to block the shipments and the governor is suing the Energy Department. Dumping this weapons-grade plutonium on our site turns us into a terrorist target. "We cannot allow the federal government to paint a bull's-eye on South Carolina," Hodges, a Democrat up for reelection in November, said earlier this year.

With its cheap labor and little concern about the environment, South Carolina has long been home to some of the nation's most dangerous substances. In addition to the Savannah River Site, the state has a low-level nuclear waste dump in Barnwell. The hazardous waste landfill near Sumter and a medical waste incinerator in Hampton have been shut down in recent years by state officials, reflecting what some see as greater environmental awareness.

The Savannah River Site is about 20 miles east of the Augusta, GA area with 477, 000 people and 170 nukes east of Atlanta, a metropolitan area with a population of 4 million.

In 1950, the government bought up 300 square miles of land near Ellenton and over the next three years constructed the five Savannah River reactors that would be used to process plutonium for nuclear weapons. During the height of the Cold War, Savannah River employed 26,000 people. Now with the Cold War over, about 13,000 people work there cleaning up the leftover nuclear material and getting it ready to ship to New Mexico to be stored underground permanently.

Underground Waste Storage Tanks

In 1998 the EPA requested that each state identify the major sources that potentially threaten groundwater in their state. Nearly three-fourths, 37 states, rated underground storage tanks as the most serious threats to their groundwater quality. Septic systems, landfills, industrial facilities, agriculture, and pesticides were also important contamination sources.

The National Well Water Association and the USGS said, about 1 percent of all United States groundwater sources which provide drinking water to half the nation, has been contaminated. They generally find groundwater contamination in areas of dense population and industrial activity. It is clear the problem is getting worse.

Nearly 40 percent of all groundwater contamination starts from leaking underground storage tanks. They are used to store hazardous and toxic chemicals and diluted waste, but an estimated 1.5 to 2 million underground storage tanks are used to store gasoline. Most of those are located beneath the many thousands of service stations in the United States.

Steel tanks eventually rust and disintegrate releasing their contents into the ground, and an estimated 84 percent of the tanks are made of steel. The steel tanks were replaced by fiberglass tanks, but they also leaked. If tanks disintegrated, why were they used?

Underground storage tanks leaking were identified as the leading source of groundwater contamination by the Environmental Protection Agency in both of its 1996 and 1998 reports to Congress about national water quality. In general, most underground storage tanks are found at commercial and industrial facilities in more heavily developed urban and suburban areas. They are used to store gasoline, hazardous and toxic chemicals, and diluted wastes. Gasoline leaking from tank systems at service stations is one of the most common causes of groundwater contamination. The primary causes of leakage are faulty installation, corrosion of tanks and pipelines.

One gallon of gasoline can contaminate one million gallons of water, or the amount of water needed for a community of 50,000 people. MTBE, a gasoline additive, is particularly troublesome because it migrates quickly through soils into ground water, and very small amounts can render groundwater undrinkable.

In 1988 the EPA issued comprehensive and stringent rules that required devices to detect leaks, modification of tanks to prevent corrosion, regular monitoring, and immediate cleanup of leaks and spills. By December 1998 existing tanks had to be upgraded to meet those standards, and replaced with new tanks of durable non-corrosive materials or closed. As of February 1999 about 386,000 releases of contaminants from corroded underground storage tanks had been confirmed. The EPA estimated that about half of those releases reached ground water. Cleanups were underway at more than 302,000 sites, and more than 192,000 cleanups had been completed.

Underground storage tank owners and operators had to meet financial responsibility requirements that ensured they would have resources to pay for costs associated with cleaning up releases and compensating third parties. Many states have provided financial assurance funds to help owners meet the financial requirements. In about 95 percent of the cases, the EPA or the states have succeeded in getting responsible parties to perform the cleanups.

In May 2001, the Government Accounting Office released its report that improved inspections and enforcement would better ensure the safety of Underground Storage Tanks, UST's. The report showed that by September 2000, approximately 1.5 million uses, underground storage tanks, had been closed permanently, leaving an estimated 693,107 tanks subject to federal regulation. To monitor this large amount of tanks, the EPA elicited the states assistance in implementing and enforcing the UST program.

Underground Waste Injection Wells and Public Wells

There are over 261,000 underground injection wells in this country which a wide range of chemical wastes are being disposed. The practice is based on the assumption that waste can be safely injected into deep, confining geologic strata or saline aquifers with no intended uses, but existing knowledge of the implications of underground injection is scanty. Ground water contamination has occurred near injection wells, and shifting within the geologic strata has the potential of freeing confined wastes into the surrounding groundwater.

The SCWA was established to prevent groundwater contamination by underground injection wells, and is also keyed to the Primary Drinking Water Regulations. Migration of fluids from injection wells into underground sources of drinking water is permitted "unless the presence of the contaminant may cause a violation of any primary drinking water regulation or may otherwise adversely affect the health of persons." This provision permits the leaching of injected liquid wastes into underground drinking water supplies until monitoring detects contamination levels that will violate one of the short lists. Hopefully no persons are dead or sick by that time.

Within the last few years, public wells have been closed in 22 communities in Massachusetts, 29 in Connecticut, 25 in Pennsylvania, and 60 in New York. Over 100 private wells have been closed in New Jersey and 500 in Long Island. In the San Gabriel Valley in California, 107 of 349 public wells serving 4.5 million people have significant levels of trichloroethylene, an organic chemical known to cause cancer in mice. Also a 1983 CRS summary of contaminated drinking water wells indicated that 2,820 wells have been either affected or closed as a result of toxic substance contamination in the last few years. There is also the threat to groundwater from abandoned hazardous waste sites.

Pesticides, solvents, and other toxic chemicals forced closure of wells used by millions of people from Maine to Hawaii. Radioactivity from mine tailings or atomic-waste disposal contaminates groundwater. "Eight million people are potentially exposed to contamination of their water supply by leakage from hazardous waste sites."

Incinerators

Incinerators are very expensive to build, but America is always putting a dollar value on something and not considering the value of the dollar spent. The country's largest incinerator in Detroit cost $438 million. It produces enough steam to heat half of Detroit's central business district and enough electricity to supply 40,000 homes. Most experts agree that energy recovered from Municipal Solid Waste has the potential for making only a limited contribution to the nation's overall energy production, but provides a conduction of health that cannot be measured.

Trucks dump waste into a large organized pit in the ground. The waste is moved to the furnace next to the pit by a crane and the furnace burns the waste at a very high temperature. It heats a boiler that produces steam for generating electricity and heat. As the garbage burns, ash collects at the bottom of the furnace where it is later removed and taken to a landfill to be thrown out. It produces poisonous gases, primarily dioxin and mercury, which are increasingly being found to be dangerous.

In the process of burning paints, fluorescent lights, batteries or electronics, mercury is released as a gaseous vapor known to be poisonous to humans and to the environment. In China where electronic components are being salvaged, there is a large health problem.

Health and air contamination problems abound with incinerators burning garbage. It gives off tiny amounts of dioxins and furan, two very dangerous chemicals, and the ash left after the burning also contains dangerous waste (furan is a flammable liquid that is obtained from wood oils of pines or can be made synthetically) (It is used in manufacturing of nylon). An incinerator that burns 1,000 tons of trash per day can generate between 200 to 250 tones of ash as residue. New regulations require that scrubbers, which pull the pollutants from the smoke, be put on smokestacks to catch the dioxins and furans. The environmental Protection Agency and Congress have not yet decided how to handle the ash. This is similar to not knowing where to store nuclear waste or what to do with it.

Dioxin is the common name for a family of chemicals, approximately seventy-five in number, with smaller properties and toxicity. Dioxins are not deliberately manufactured. They are the unintended bio-products of industrial processes that involve chlorine or a process that burn chlorine with organic matter. Incinerators provide an ideal environment for dioxin production, with plenty of chlorine sources and precursors to dioxin combined with sufficient heat to drive the reactions that produce dioxin.

It is a stable compound that accumulates in the human body over a lifetime. The EPA calculates the average level of dioxin in the body of a middle aged person to be 9 monograms per kilogram.

Andrew D. Anderson

Chapter 5: Our Trees Are Dying

Insect and Wildlife Problems

Rats, ants, wasps, snakes, squirrels, and skunks enjoyed the balmy days so much that they spent much of their time procreating instead of coping with realities of a normal New England winter in 2002. They are emerging and are one of the large west explosions of urban wildlife in recent memory, according to wildlife specialists and pest control companies.

It is going to be a very busy pest season, said Victor Palermo of Ultra Safe Pest Management in Quincy, MA. Calls from homeowners and businesses with insect and wildlife problems are double what they were at that time last year. With not much snow and ice to make food scarce and only a few stretches of severe cold weather to cause the normal level of winter kill, various forms of urban-dwelling instead survived the winter with relative ease.

Raccoon, squirrels, and skunks are having litters early and, in some cases, may have had at least one extra litter this year, said exterminators. Bats, which normally hibernate, were active for most of the winter, awakened by spring-like weather or conditions to mate. Rats, which normally are sexually dormant during frigid weather, reproduced as though it was spring, having multiple litters making the rat problem worse. Female pups born this winter may be already mature enough to have litters of their own.

I expect very sharp increases in rat populations all over New England, said Bruice Colvin, a Lynnfield ecologist who is considered one of the world's leading authorities on rats. When you have a mild winter, there is more food and more warm places for them to live, with less stress. Instead of killing each other in the competition for food, they are all getting along and reproducing.

Large colonies of insects such as carpenter ants, which bore into wood to make their nests as well as unusual numbers of bees, wasps, and other winged bugs, have also appeared months before their normal arrival around June.

Coming at a time when signs of global climate changes have begun to persuade more scientists that the earth is warming, some now speculate that such population explosions are more than an occasional change in the balance of nature. More and more things are beginning to add up, said Thomas Peterson, a research meteorologist at the National Climatic Data Center in Asheville, NC. There is a lot of evidence that some of these changes are real and that they are affecting ecosystems.

A recent Cambridge University publication concluded that shifts in the timing of events such as breeding to early in the spring already have had profound effects on animal populations around the world. Boston city workers have been laying out bait and traps while also cracking down on overflowing garbage dumpsters around the city. Home Depot stores in Boston report such high sales of rat traps and various pest baits and poisons that they are making more room on the shelves to meet the demand.

Animal welfare groups are offering tips on how to keep rats and other animals at bay. Homeowners should make sure the kitchen is clean and the garbage is properly stowed, said Stephanie Hagopian, director of the Living with Wildlife program. They should fill any holes in the foundation of their houses with cement and stuff copper material, one of the few substances rats won't gnaw through, into interior gaps. These animals have been here for hundreds of years and are not going anywhere. So we have to learn to live with them whether we like it or not.

Destroying Our Hemlock Trees

A tiny relentless insect is decimating the big Hemlock trees at Burnham Brook, CT. The trees once grew so thick and close to each other that other plants could not survive in their shade. They were masters of the 450 acre woods, creating a cool open world beneath their boughs where visitors had escaped the summer sun for generations.

Today the forest looks like it was ravage by a forest fire. Seventy-foot hemlocks stand bare as far as the eye can see, surrounded by their own fallen branches basking in the sunlight that finally streams through the trees. This is the unfortunate handiwork of the woolly Adelaide, a bug no bigger than the period at the end of a sentence, which is decimating some of the Northeast's oldest and most valuable forests. They lack in size but make up in numbers, swarming the tiniest hemlocks by the tens of thousands and literally sucking the life out of them.

Looking at a six-inch hemlock seedling at Burnham Brook coated by the Adelaide's white egg sacs, Dave Orwig an ecologist of Harvard University sees a species that has been and important part of evergreen forests from Maine to Virginia for 8,000 years. It could drive out other trees until all that remained were the dense green foliage of the hemlocks, creating a shadowy, primal forest floor.

They first arrived in the Pacific Northwest from Japan around 1921 perhaps in a shipment of ornamental trees. They seemed to withstand the bugs just fine even when the Adelaide was inadvertently transported to Richmond, Virginia in the 1950s.

Today most old growth forests in Massachusetts from Mount Greylock in Adams to the Virginia Woods in Stoneham are dominated by hemlocks. Mature hemlocks also played an important role in preventing soil erosion around water bodies, since they grow in moist soil, and they are logged for everything from lumber to high-grade mulch.

Just 14 years after their arrival in New England, the quiet insect aggressors have reached every mainland county in southern New England, invading everywhere from the woods that surround Greater Boston's water supply to the trees that shade the Brookline golf course where the Eider Cup Golf Tournament was played. No one has found a way to stop the bugs 12 to 18 mile-a-year northward advance from Virginia that has already covered 11 states.

The speed and thoroughness of the Adelaide invasion draws comparisons to other infamous tree scourges such as Dutch elm disease and the fungus born affliction that killed many of the graceful shade trees that lined New England streets over the last 50 years. Like the fungus, originally from Asia that causes the elm disease, the woolly Adelaide comes originally from Japan and has no natural enemies to control its population here. In its native Japan, the Adelaide poses little threat to overall hemlock health.

Acid Rain

. Acid rain has become so bad in some northeastern states that all fish and other animal life in the lakes have been killed, and air planes dump lime on the ice during the winter to neutralize the water once melting occurs.

Many lakes in Connecticut, Southern Massachusetts, Rhode Island, the Adirondack Mountains of New York, and the Pocono-Mountains of Eastern Pennsylvania have lost sizable populations of fish. Many species of freshwater fish have become extinct since 1979 and many additional species have become endangered.

President Bush asked Congress to approve mandatory limits on total industry output of three kinds of pollutants, and to let companies work out how to achieve them through a system of earning and trading credits. The pollutants are smog-causing nitrogen oxide, acid rain-causing sulfur dioxide, and mercury, a toxic chemical that contaminates waterways and goes up the food chain through fish to people.

Acid rain is caused when pollutants are carried east on winds from Midwest smokestacks and mix with water vapor in clouds over the Adirondacks. The Adirondack Council estimates that 500 to 700 of the Adirondacks' 2,800 lakes are too acidic for native fish and plants. Also many high-elevation spruce trees have been damaged and some strains of trout have disappeared. It is believed that by 2040 half of the life in the Adirondacks lakes and ponds could be deserted by acid rain.

Another problem we have with water is acid rain that is a result of pollution from manufacturer, automobiles using fossil fuel, and manufacturers that burns coal. Automobiles emit sulfur dioxide and nitrogen oxides and when they combine with moisture in the atmosphere, they form sulfuric acid and nitric that falls to the earth as fog, acidic rain, snow or dust. Before we continued to generate more of these chemicals than required, we had a balance nourished aquatic and plant life. By 2010, power plants are required to reduce their sulfur dioxide emissions to 50 percent of what the 1980 level was. When you look at that statement, you realize we will never stop the emissions. The best we can hope to do is reduce emissions over a longer period of time, but it will never be enough.

When spring snow melts there is an accumulation of snow and acid over the winter months and the runoff empties into the lakes or rivers. It affects our forests, many things that lives there, kills our separate trees, and corrodes our car's paint. About 60 percent of the sulfur dioxide comes from power plants that rely on burning fossil fuels such as coal to generate electricity. This problem is apparent more in the Northeast because the region is downwind from Midwest power plants that release high levels of sulfur dioxide.

They also produce nitrogen oxide that is a contributor to acid rain and motor vehicles also emit high levels. That acid rain falls in streams, rivers, and lakes making them more acidic for fish.

At high elevations the thin soil becomes acidic from the pollutants and dissolves nutrients that trees need to thrive on. The authors of the articles estimated 25 percent of the red spruces in the White Mountains have been wiped out because of acid rain, and forest and trees have not recovered because more acid is being put into the air every day. The trees once had some help as mineral buffers in the soil from calcium that is suppose to neutralize the acid, but it has been depleted, thus the trees have no help. The trees do not get much calcium so they are susceptible to insects infestations and weather change says Charles T. Driscoll and Gene E. Liken, one of the report's authors and director of the institute of Ecosystem Studies in Millbrook, NY.

Acid rain was first identified in North America in 1972 in the Hubbard Brook Experimental Forest in New Hampshire. The recent study also took place there. Acid rain continues to have a significant ecological effect, he said. Even now, forests and lakes are not recovering from the toxic effects of acid rain despite significant cuts in the power plant emissions that cause it. It also causes more environment damage than projected a decade ago, the researchers reported in the journal Bio Science. The teams say an additional 80 percent reduction is needed to bring sensitive streams back to non-acidic levels with in 25 years from 2001.

The effect of acid rain on trees adapt to environmental stress better than others. It depends on the type of tree, its height, and its leaf structure on how well it will adapt to acid rain. Acid rain may affect trees in at least two ways. In areas with high evaporation rates, acids will concentrate on leaf surfaces. In regions where a dense leaf canopy does not exist, more acid may seep into the earth to affect the soil at the base of the tree.

Scientists have done a study on the effects of acid fog on trees. In fog, the concentration of acid and sulfate in the droplets is much greater than in rainfall. In areas of frequent fog as in London, damage has occurred to trees and other vegetation because the fog condenses directly on the leaves.

A joint report of the European Commission and the United Nations Economic Commission for Europe in 1994, studied 102,300 trees at 26,000 sampling plots in 35 European countries and found that almost one-quarter of the trees in Europe were defoliated by more than 25 percent. Forest damage is a problem in all European countries. Fifty three percent of all trees in Czech Republic suffered moderate or severe defoliation or have died.

Acid precipitation cause leafy plants, such as lettuce, to hold increased amounts of potentially toxic substances, such as the mineral cadmium. It has effects on Statues and Monuments. The problem was particularly apparent in the Northeast because the region is downwind from Midwest power plants that release high levels of sulfur dioxide. The winds travel from west to east and from south to north. Fifteen percent of lakes in New England are either chronically or periodically acidic.

The Capacity to Absorb Acid

The news is worrisome to conservationists who fear that President Bush, who recently reversed a campaign pledge to regulate carbon dioxide from power plants, will be unwilling to place any new limits on electric utilities. About 60 percent of the sulfur dioxide found in the atmosphere comes from power plants that rely on burning fossil fuels, such as coal, to generate electricity.

Acid rain leaches calcium out of the soil and robs snails of the calcium they need to form shells. Because mice and species of songbirds get most of their calcium from the shells of snails, the birds are perishing. The eggs they lay are defective, thin and fragile.

Although all soils have some capacity to absorb acids, just how much depends on the chemistry and thickness of the soil in question. Some soils have a protective mechanism, calcium carbonate, these are the limestone soils. The protective mechanism works on the chemical principle that acids and bases neutralize each other when mixed together. The limestone in the soil reacts with the acid in the rainfall, neutralizing the acidity, thereby lending protection to the lake water. By the time the rain has filtered through the soil and reached the lake, the acidity has been neutralized. Unfortunately, few if any of the Adirondack's 2,500 lakes possess this protective chemistry. At last count, the Adirondack Mountain Region in the Northeastern United States has 200 lakes which are to acidic to support life and another 200 which show some damage.

Canada, to the north, has 2,000 dead lakes, and if the amount of air pollution from both Canada and the United States are not sharply reduced, another 48,000 lakes will be threatened over the next 20 years. Presently, the United States produces about five times more sulfur pollution than Canada, sending about 2 million tons of sulfur dioxide per year on the prevailing winds.

Acid Problems in Other Countries

In Sweden, there are at least 3 to 4 thousand severely acidic lakes and another 14 to 15 thousand which are damaged. Out of its 90,000 lakes, 18 to 20 thousand of them are affected by acid rain, numbers which indicate and advanced stage of ecological decline. Although both Sweden and Norway on the Scandinavian peninsula have adopted strict measures to reduce sulfur emissions from their own industries, their acid rain problem persists because the major portion of the acid rain affecting their environments originates in Great Britain, France, and Germany. Sweden is responsible for only about 22 percent of its sulfur burden, and Norway contributes almost 20 percent of its sulfur pollution. Europe's sulfur emissions remained at about 25 million tons yearly until 1950. Since 1950 this level more than doubled reaching the level of 60 million tons yearly in 1973 which would add millions of tons in 2002.

The national Academy of Sciences recently endured the conclusions of work by the Environmental Defense Fund's staff physicist, Dr. Milchael Oppenheimer, providing that acid rain can be reduced by curbing the emissions of coal-fired power plants. Nationwide, American industry puts about 26 million tones of sulfur dioxide into the air each year. Coal-fired power plants contribute nearly two-thirds of this total.

Acid rain falls everywhere, on cities, on forests, and into lakes. When it lands in the cities, it can damage stone statues and buildings. When it is in smog, it can harm people's lungs. When it lands on the forests, it can kill trees. When it lands in lakes it acidifies the water. That in turn, harms the plants and animals living in the lake. Small increases in the acid level in the lake can make it harder for the fish to reproduce, and higher levels of acid can kill all the fish in the lake.

Many states add limestone or other alkaline materials to neutralize the acidic affect on the lakes, although this often works for only a short time.

Mining companies have to separate the stone and dirt that they take from mines to get the mineral ore that they want. To do this they use chemicals, many of which are acidic. Sometimes the waste from this method of extracting ore end up in rivers or lakes and when this happens it makes the lake to acidic and kills the plants and fish in the lake.

Lake depth, biological life, oxygen levels, and the inherent clearness of the water are natural aging process and are called eutrophication. Lakes are clear waters with little organic matter or sediment. Mesotrophic waters contain more organic material, and the oxygen level is being depleted: Eutrophic water is murky and shallow with lots of algae and a depleted oxygen level.

Hundreds of lakes in the Northeast United States are already clean of fish by acid rain, along with hundreds of miles of streams. In many lakes restocking may be impossible because of permanent changes in the chemistry of the water and surrounding soil. Tens of thousands of lakes and as many miles of streams are endangered, many already showing alarming decreases in their capacity to buffer the acid falling into them from the sky.

Although the sulfur content of acid rain has decreased 38 percent since controls were put into effect, the forests and trees have not recovered as expected the study finds. A mineral buffer in the soil such as calcium, which neutralizes the acid has been depleted. With that buffer gone, upper levels of soil are that much more sensitive to acid rain. The forest have been bathed and rebathed by acids for many years and it continues to get worst.

Trees Dying from too Much Acid

Trees are dead or dying on 450,000 acres and the spruce forests around Rybnik. Czestochowa, the southern industrial region, is completely gone. Environmentalists have warned Poland that as many as 7,500,000 forest acres were destroyed by 1990, based on the burning of high-sulfur brown coal.

In the United States, forest damage is most evident in the Appalachian mountain ranges in the east and in the Sierra Nevada in California. It has documented not only tree diseases and destruction of the forest but also sustained declines in growth. Damage is most severe in the high-elevation forests of New York, Vermont, and New Hampshire. Of those three states, the greatest documentation of forest decline is on Camel's Hump in the Green Mountains of Vermont. Between 1965 and 1979 researchers found that seedling production and tree density had declined by about half, but in 1979 over half of the spruce trees on Camel's Hump were dead. The blame is on acid rain.

Some scientists have firmly documented that tree disease is liked to the availability of ozone and other pollutants in the family of photochemical oxidants. Ozone forms when nitric oxides react with hydrocarbons in the presence of sunlight, and a long time study of the effect of ozone on the pine forests in the San Bernardino Mountains east of Los Angeles reveal that the yellow brown photochemical smog had a devastating effect on the trees.

Acid rain has an adverse effect on naturally acidic soils because of the behavior of ions in the soils as their acidity is increased. In other areas scientists have firmly documented that the disease is linked to the availability of ozone and other pollutants in the family of photochemical oxidants.

The Brookhaven National Laboratory, and the Army Corps of Engineering in 1993 found that acid rain caused "$5 billion" worth of damage annually in a 17 state region. Two thirds of the damage was created by pollution whose source was only 30 miles away.

Every third tree in West Germany's forests is damaged. About 18,000 Swedish lakes have become virtual acid pools. In Poland, two-thirds of the rivers are so polluted the water is not fit even for industrial use without purification. Czechoslovakia, the forests in the huge Erzgebirge Mountain range are rapidly dying. An area of about 247,000 acres lies bare, the vegetation killed by contaminants produced by Czech power plants to the south.

Savant Oden's rain maps clearly showed the growing acidity spreading out from pollution sources in Europe's industrial center. Lake chemistry, which is always closely linked to the land, is changed by the action of acid rain. The Eastern portion of the United States receives rain which is 10 times more acidic than what is considered normal. This acid rain dissolves aluminum from the soils surrounding a lake and enters the lake water. Fish are vulnerable to this combination of aluminum and acid, their gills erode, become clogged and the fish die from suffocation. Dissolved aluminum also kills mycorrhiza fungi which is important to the health of most tree species.

Acid rain and the pollutants that cause it can be a health hazard to children, causing asthma and bronchitis. It is also suspected of health risks for those over age 65, chronic bronchitis, asthma, emphysema, pregnant women, and those with histories of heart disease. Dr. Philip Landrigan, director of environmental and occupational medicine at the Mount Sinai School of Medicine says acid rain is probably third after active smoking and passive smoking as a cause of lung disease.

Transportation System for Acid

Wind in the atmosphere carries the chemicals in the direction it is blowing. Japan's rain is increasingly acid, because of China's big dirty coal-burning factories. A study released in 1994, said China is responsible for half of Japan's acid rain and Korea for another 15 percent. It will be almost impossible to control the pollution because China is the world's biggest consumer of coal and is growing.

In the spring and winter, large air masses from continental Asia move to Japan propelled by prevailing monsoons causing the acidity levels to be the highest. Associated with the high levels, their forests in Japan have a high death rate of red pine and Japanese cedar.

Europe has a transport pattern that carries pollutants from the United Kingdom over Sweden to Southwestern Germany and half of the trees of the famed Black forest are dying from the effects of acid rain.

There is a transport system that distributes acid emissions in exact patterns across the planet. A typical transport pattern in the United States is from the Ohio River Valley to the Northeastern United States and Southeastern Canada.

The climate contributes to the movements. In humid climates as along the eastern seaboard where there is less dust, precipitation is more acidic. But in drier climates as Western United States, windblown alkaline dust blows freely and tends to neutralize atmospheric activity. In the season of the year when fish are spawning or seed is germinating is the most harmful period.

Pollutants are concentrated in upper layers of snow packs in cold climates. When it thaws or rain falls on the snow, the concentrated acid is washed into streams and lakes. The mountain streams in North Carolina, Tennessee, Pennsylvania, New York, and Arkansas have shown increased acidity between 3 to 20 times the level the rest of the year. Since many fish hatch in the spring, it can harm or kill new life and also kill the adult fish. The increase affects insects, snails, and crayfish which other fish feed on.

Dust Clouds That Travel

Drifting with the suspended dust particles are soil pollutants such as herbicides and pesticides and a significant number of microorganisms, bacteria, viruses, and fungi. Using a quick calculation, assuming there are only 10,000 bacteria per gram of airborne sediment and assuming a modest one billion metric tons of sediment in the atmosphere, the numbers will translate into a quintillion (10 to the eighteenth) sediment-borne bacteria moving around the planet each year. It would be enough to form a microbial bridge between Earth and Jupiter.

It has been estimated that 13 million metric tons of African sediment fall on the North Amazon Basin of South America every year. Dust storms originating in North Africa routinely affect the air quality in Europe and the Middle East. Reports of a fine red layer of African dust on automobiles and snow are not uncommon in Western Europe.

During the African dust events, microbes present in the Caribbean air has been measured and identified. Twenty five percent were species of bacteria or fungi that have been identified as plant pathogens and about 10 percent were opportunistic human pathogens, organisms that can infect people who have a lowered resistance. This gives good evidence that the fallout has had direct consequences on the health of coral reef communities in the Caribbean and human health.

A giant dust storm above Northwestern Africa blew out into the Atlantic over the Canary Island on February 11, 2000. Dust storms like this one carried infectious microbes and toxic chemicals and were believed to have a significant hazard to ecosystems in the Caribbean and other parts of the Americas.

In the 1990s, satellite images revealed the magnitude by which desert soils are aerosolized into giant clouds of dust. The energy for this massive launch into the atmosphere comes from rapidly moving high-pressure systems and storms that produce powerful winds. Once carried aloft, the sediments and their tiny inhabitants are swept along by the atmospheric circulation patterns, often settling many thousands of kilometers from their site of origin.

The Sahara desert covers the Republic of Mali's northern half, and the semi desert lands of the Sahel stretch south toward Guinea and Burkina Faso. Mali is economically poor and basic sanitary facilities are lacking. The Niger River flows through thousands of kilometers of Mali's arid lands and is the repository for all types of waste, including animal feces and excreted pharmaceuticals that are used against a broad spectrum of human diseases. Once a year, the river deposits a load of fine sediment on the flood plain, along with whatever else it carries. People plant crops on the new deposited soil and burn garbage to fertilize the soil.

The garbage also is of animal and plant waste, plastic products, and rubber tires. Small fires burn day and night creating black smoke and releasing plasticizers, polyaromatic hydrocarbons dioxin, and heavy metals. The combustion products are adsorbed in the clay soil, which is then injected into the atmosphere by strong winds. The small particles absorb chemicals from the atmosphere, other pesticides, combustion products and cosmogencally produced radioactive isotopes.

Some of the dust particles are so small that once they are inhaled into the lungs they cannot be exhaled. Anything present in the dust is also carried deep into the lungs, close to the capillary beds. Some of the contaminants are endocrine disrupters (pesticides and polyaromatic hydrocarbons), some are carcinogens (dioxin and radioactive isotopes) and others are simply toxic to cells (heavy metals).

The Aral Sea had a surface area of about 60,000 square kilometers in 1960. It is now less than half its original size because of the diversion of source waters for agricultural purposes. The formation of dust clouds over its seabed during a storm is common, and high concentrations of pesticides and herbicides have been reported in airborne sediments. Exposure to these regional dust storms has resulted in illness and hospitalization, and DDT residues have been found in human breast milk.

An enormous swirling dust cloud that originated in the deserts of China and Mongolia, soared across the Pacific Ocean to Western Canada, raising fears among meteorologists that it could endanger aircraft. It was tracked by satellite and estimated it stretched for 1,250 miles.

The massive dust cloud began as a sandstorm in the Taklimakan Desert in Western China and the Gobi desert in Eastern Mongolia in early April and was lifted by heavy winds said Servranckx in 2001. Picking up industrial pollution over China, it crossed Japan and Russia to the Gulf of Alaska, and came down in Canada. They usually do not travel long distances.

Weather experts say the cloud probably made it that far because it caught a ride on some very high air flows traveling as much as 7 miles above the earth. It almost looked like a cloud, but it was darker and did not have exactly the same color.

Clumps of the cloud still hover over parts of the Canadian Prairies and the Great lakes and weather forecasters expect it to soon drift into the Midwestern and Eastern United States. All of the low level dust has fallen out, so it is caught in the jet stream said Jay Anderson a meteorologist with Environment Canada in Manitoba. Meteorologists said the cloud was a dramatic demonstration of the movement of the atmosphere. Long-range transportation makes the problem of abatement a particularly difficult one because ozone may be carried by wind currents great distances from its area of origin.

In the recent years most of the pollution publicity has been concerned with the quality of the outside air, but numerous investigations have revealed that the air inside buildings is contaminated with a large variety of pollutants as well. They range from natural contaminates, such as radon, to contaminants that are the result of human activities, smoke, and combustion by-products.

Virus and disease causing particles also travel on the wind. New England could be losing one of their best defenses against disease, cold weather. Every fall mosquitoes that may be carrying the deadly West Nile virus, are killed off before they multiply and spread the disease too widely. But as global warming heats up the earth, by minute degrees, disease-carrying organisms may regenerate faster or go into new areas where populations may have little or no natural resistance.

The report notes that with increased temperature, mosquitoes that carry and dengue virus bites more often. Slime mold grows faster and parasites that attach to butterflies gather in greater density. The spread of Rift Valley fever and even eastern lobster disease appears to be largely related to long-term temperature fluctuations. Rift Valley is a nasal disease that can make victims go blind and vomit blood. It spread across the Red Seas by winds in 2000 and killed 200 people in Yemen and Saudi Arabia and its range is expanding now in Maine waters that have warmed slightly in recent years. While there are multiple reasons for the redistribution of emerging disease, it is clear there is an emerging pattern here, said Paul Epstein, Associate director of Harvard Medical School's Center for Health and Global Environment.

There is other evidence that bacteria, fungal spores, and viruses may spend large amounts of time in the air riding clouds across the planet. Since there is mention of harmful rays coming through the clouds, Gene Shinn of the U.S. Geological Survey in St. Petersburg, FL, says the bacteria seem to protect themselves from harmful rays by attaching to dust particles. In dust clouds, the amount of Ultra Violet Radiation will be lower than in normal situations. Dale Griffin said that bacteria might survive even longer if they get into cracks in the particles. They can then travel longer distances and spread disease on arrival. He also says bacteria and fungi carried aloft on dust storms coming out of the Sahel region of West Africa can journey across the Atlantic in large numbers. He has isolated more than 130 species of African bacteria and fungal spores over the Caribbean.

Griffin suggested the 2001 Bristish epidemic of foot and mouth disease may have arrived in Europe on winds from Africa. The first case was reported just a week after satellite pictures had shown a huge dust storm carrying sand from the Sahara to Britain. To reinforce this, a previous outbreak of foot and mouth in Britain was traced to the virus blowing across the English Channel from France.

What The Future Holds

Burning of coal is projected to be increased as a fuel causing an increase in sulfur dioxide emissions and more damage to aquatic animals and plants. America has never realized that once the soil and waters have been acidified, a return to normal will not exists causing a continuous lost in animal species, agricultural productivity, and forests.

The availability of ozone in the atmosphere is critical to the maintenance of life on earth. Although small quantities of ozone may occur close to the earth in the troposphere, it is concentrated in the stratosphere from 30 to 60 miles above the earth.

Federal legislation has recently been filed to cut sulfur emissions and additional 50 percent. But scientists say reductions need to be made as soon as possible if the Northeast is to recover anytime soon.

Acid rain can be reduced by curbing the emissions of coal-fired power plants. American industry puts about 26 million tons of sulfur dioxide into the air each year. Coal-fired power plants contribute nearly two thirds of this total. Hundreds of lakes in the Northeast United States are already wiped clean of fish by acid rain, along with hundreds of miles of streams. In many, lake restocking may be impossible because of permanent changes in the chemistry of the water and surrounding soil. Tens of thousands of lakes and as many miles of streams are endangered, many already showing alarming decrease in their capacity to buffer the acid falling into them from the sky.

Andrew D. Anderson

Chapter 6: Changes Required

Progressive Changes

With 6.1 billion people relying on the resources of this same small planet, we are beginning to realize we are drawing from a finite account and making no deposits. The amount of crops, water, and other bio-matter we extract from the earth each year is exceeding what the planet is programmed to replace. Meaning it would take just a little over 14 months for the earth to replenish what we selfishly use in 12. It is and always will be a challenge for us to live in harmony with the earth as it is a challenge to live with other humans on this earth.

Exiting the 19th century, the internal combustion engines, the steam engine, the industrial revolution, the energy locked in the fossil fuels like coal, gas, and oil, we exploited on a large scale. Society also exploited the Indian, the land, the buffalos, the slaves, and societies self defined freedom.

It is imparities that we need another revolution of global stewardship. Society believes our stewardship is based on our freedom to exploit anything we choose as long as we have a desire for it. But what it should mean to be citizens of this planet is accepting our obligations to be stewards of the earth life-giving capacities.

Since burning coal causes a large portion of the dirty air we breathe, lets find a new equation that does not include coal and the answer is in the direction of clean air.

Free Renewable Energy

Over the Columbia River on a high desert ridge, the world's largest wind farm sprawls across 50 square miles of Oregon and the state of Washington. When the last of its 460 turbines are installed, this post modern power plant will offer clean electricity to 70,000 homes and businesses. Every month hundreds of tourists stare at its fiber-glass blades twirling with ballet grace atop 160-ft. poles. People are in awe of wind power, says Anne Walsh, community relations manager of the Stateline Energy Center.

Wind has been with us since the beginning of time and is becoming more than a sideshow behind the curtain of the world. It is now the world's fastest growing power source, a high-tech challenge to the coal mines, oil rigs, nuclear reactors, and hydroelectric dams that goes with the 20th century. Experts say wind could provide up to 12 percent of the earth's electricity within two decades.

The author grew up on a farm in the state of Kansas and we had electricity at night in the 1930s from a single bladed propeller attached to a generator that generated electricity by rotation. When there was a storm the lights were bright as the rpm increased. Cape Cod in Massachusetts is beginning to use wind power that is available for use. What better source of unpolluted power with no odor, no noise, and you cannot even see the wind.

One large renewable-energy proposed in the United States is 170 wind turbines that will generated 420 megawatts of power located on Cape Cod in Massachusetts. It will be enough electricity for a city of 200,000 people. The developer is Cape Wind Associates and is scheduled to start construction of a 200-foot tall test data tower this year in 2002.

The turbines will be 250 feet high and the blades will reach another 170 feet above the hub and they usually rotate 16 times per minute and will be located in rows. It should replace enough power-plant electricity annually to eliminate 4,600 tons of sulfur dioxide, 1,566 tons of nitrous oxide, and 120 tons of carbon monoxide that will reduce the greenhouse emissions more than a million tons per year.

Well financed energy have proposed major wind farms in New England that would bring 270 towering wind turbines to the mountain tops and coastal waters and could generate enough electricity for as many as 250,000 homes. The efficiency of turbines is making wind a serious competitor with fossil fuels, leading wind hopefuls to spend tens of millions of dollars on proposals.

Catamount Energy of Rutland wants to build a $50 million generating plant with as many as 27 200-foot-tall turbines stationed along a ridge of Glebe Mountain over looking a Vermont hamlet. Florida Light and Power plans to increase wind generation in 2003 that is estimated to cost $4.7 billion when completed. The people are having debates over the merits of altering gorgeous views in exchange for pollution-free electricity.

Selectwoman Claire Trask says Glebe Mountain looks more pristine than it is. There are farms and stonewalls, roads, and power lines. And on the north end of the proposed site is the Maic Mountain ski area with its trails and ski lifts cutting through the evergreen face of the mountain. A 29-turbin, 52-megawatt facility in Redington Township, Maine is also planned. In the Western part of America, farmers get $2,000 annually for each turbine sited by a power company on their land.

Wind farms in Texas, Oregon, Kansas, California, Arizona, and elsewhere helped lift the United States wind-energy output 66 percent last year in 2001 and an additional $3 billion in American projects are in the works. Wind is competitive, wrote Mark Moody-Stewart, the former chairman of Royal Du Shell who now co-chairs an alternative-energy task force for the Group of Eight.

Ireland is preparing construction on the first of 200 wind turbines that will generate 1,000 megawatts of electricity. The United States has 21 windmill projects proposed from Maine to Carolina stated John P. DeFillars the executive vice president of Brownfield's Recovery Corporation and a former head of the Environmental Protection Agency's Regional Office.

Windmills generate renewable power, so called because the source of the energy, wind, is continually renewed by nature, ditto for solar cells, which are powered by the sun, which uses the heat from the earth, and hydroelectricity which comes from dams. Unlike oil and coal deposits, renewable energy cannot be exhausted, at least not until the sun burns out billions of years from now and the earth goes cold. Then nothing will be needed.

Australia is using photovoltaic dishes to collect sunlight to generate electricity. In the United States, Philip Merrill Environmental Center in Annapolis, Maryland is Earth friendly also. One third of its energy comes from geothermal heat pumps that utilize the Earth's warmth and photovoltaic buildings panels that covert sunlight into electricity. The rainfall collected on the roof can be channeled into huge holding tanks for reuse in irrigation.

It is a disgrace that we have done so little to reduce our dependence on imported oil, says Reid Detchon, a former United States Energy Department official who now consults for the United Nations Foundation. Clean energy has a long ways to go. Only 2.2 percent of the world's energy comes from renewable such as small hydroelectric dams, wind, solar, and geothermal. Renewable energy from large dams provides another 2.2 percent.

Japan imports 99.7 percent of its oil to Germany, where the nearby Chernobyl accident turned the public against nuclear plants. But the momentum toward clean renewable is undeniable. Globally, solar, and wind-energy output is growing more than 30 percent annually, far faster than conventional fuels, and their cost is plummeting. We are on the edge of an energy revolution says Christopher Flavin, president of the World-Watch Institute, a Washington non-profit. It will be as profound as the one that ushered in the age of oil a century ago.

Even oil companies are trying to cash in on the de-carbonation trend. The world has gradually moved toward cleaner fuels, from wood to coal, from coal to oil and from oil to natural gas. Renewables are the next step. Japanese manufacturers, led by Sharp and Kyocera, have moved aggressively into photovoltaic cells, which turn sunlight into electricity. In April General Electric snapped up Enron Wind from the bankrupt energy giant. We are on a journey to a lower-carbon world, says Graham Baster, an executive at Britain's BP, which is building a $100 million solar plant in Spain. Solar and wind energies are intermittent, when the sky is cloudy or the breeze dies down, fossil fuel or nuclear plants must kick in to compensate. But scientists are working on better ways to store electricity from renewable sources.

Current from wind, solar or geothermal energy can be used to extract hydrogen from water molecules. In the future, hydrogen could be stored in tanks, and when energy is needed, the gas could be run through a fuel cell, a device that combines hydrogen and oxygen. The results, pollution free electricity, with water as the only by-product. Already fuel-cell buses, cars and small generators are being tested. Eventually, some visionaries say, fuel cells placed in individual buildings could replace many of today's giant electric plants. But that will not happen unless the technology is refined and the cost drops. The Apollo in the 1960's that went to the moon used three fuel cells for electricity and drinking water that the author was a senior design engineer on.

We Need Help from Government

While the developed nations debate how to fuel their power plants, some 1.6 billion people have no access to electricity or gasoline. They cannot refrigerate food or medicine, pump well water, power a tractor, make a phone call or turn on an electric light to do homework. Many spend their days collecting firewood and cow dung, burning it in primitive stoves that belch smoke into their lungs. To emerge from poverty, they need modern energy. Renewables can help, from village scale hydro power to household photovoltaic systems to bio-gas stoves that convert dung into fuel. More than a million rural homes in developing countries get electricity from solar cells. The potential is enormous, says Anil Cagraal, an energy specialist for the World Bank.

Widespread government subsidies for fossil fuels and nuclear energy, estimated at some $150 billion per year, must be dismantled to level the playing field for renewals. Policymakers must factor in the price of pollution. Coal plants are more expensive than renewable power when one includes the cost of scrubbers on smokestacks and the expense of health care for coal related illnesses. The cost of nuclear energy would soar without government insurance. Environmentalists are calling for taxes on carbon to slow the growth of fossil-fuel use.

Another way to increase the. renewable share of the energy mix is to reduce the use of conventional fuel through efficiency incentives. Experts estimate that efficiency could slash the globe's projected energy consumption by a third. Strict standards can cut energy use in everything from air conditioners to cars. Compact fluorescent lamps use a quarter of the electricity of incandescent bulbs to provide the same amount of light.

Governments are increasingly forcing utilities to make it easier for windmill and solar-panel owners to connect to the grid and get credit for providing extra electricity they do not use. Governments are also pressuring utilities to meet targets for renewable sources of energy. The European Union is requiring its members to boost electricity from renewals to 22 percent of production within the next eight years. Brazil plans to push a global standard at the World Summit on Sustainable Development in Johannesburg in 2002.

But the United States has always shunned a carbon tax John Holdren, a professor of environmental policy at Harvard's Kennedy School of Government says, such a tax could stimulate economic growth and help position the United States as a leader in energy technology. The energy technology sector is worth $300 billion a year, and it will be $500 to $600 billion by 2010, Holdren says. The companies and countries that get the biggest chunk of that will be the ones that deliver efficient, clean, inexpensive energy.

Some companies have already figured that out. One of the most advanced large corporation is chemical giant Dupont, which first acknowledged the problem to eliminate change in 1991. Throughout the past decade, the company worked to cut its carbon dioxide emissions 45 prevent from 1990 levels. Last year, it pledged to find at least 10 percent of its energy from renewable sources. The oil giant BP in 1997 agreed that climate change was indeed occurring. Even with other oil firms protesting that the evidence was too thin, BP pledged to reduce its greenhouse-gas emissions by 10 percent from 1990 levels by 2010. At the same time BP and Amoco is pouring money into natural gas exploration and investing in renewable energy like solar power and hydrogen.

Automobile manufacturing is already in a race for alternatives to fossil fuels. Several automakers like Ford, Daimler-Chrysler, and Volkswagen have developed prototypes of cars run by hydrogen fuel cells rather than gasoline.

Europe today accounts for 70 percent of the world's wind power. In Japan 80,000 households have installed solar roof panels since the government offered generous subsidies in 1994. Japan has displaced the United States as the world's leading manufacturer of photovoltaics. India established a fund that has spent $1.1 billion to alternative-energy projects. The country is now the globe's fifth largest generator of wind and solar power. Iceland, which lies on a hotbed of underground volcanic activity, uses that geothermal energy to heat 90 percent of its buildings. The island nation is planning to use geothermal and hydroelectric power to produce large amounts of hydrogen, creating the world's first hydrogen economy.

Such examples show that the future is more a matter of choice than destiny, as Brazilian physicist Jose Goldemberg, the chairman of a recent United Nations Energy study said. On the windy border of Washington and Oregon, citizen groups are already making a choice. They have pressured utilities to invest in green energy, and a federal tax credit has made it more profitable. It is the thing to do say Vito Giarrusso, manager of the Stateline Wind Farm, to help our little piece of the earth.

Andrew D. Anderson

Today a third of the earth's wind energy is produced on German soil. In 2000 another Schgeer-sponsored law increased the price for solar energy.

By mid-century, Hermann Schgeer asserted confidently, wind, solar, and other renewals can snuff out all conventional energy sources. He has transformed Europe's energy landscape. He pushed through laws that have turned Germany into the world's biggest wind-power user, surpassing the Unite States and the second biggest solar energy generator after Japan. This is remarkable for a country that is not particularly windy, or sunny. In Germany and beyond, says Reinhard Loske, energy spokesman for the nation's Green Party, Schgeer is the prime mover of the cause of renewable energy.

In 1991, he sponsored legislation opening Germany's grid to renewable-energy producers and settling a generous fixed price for their power. Today a third of the earth's wind energy is produced on German soil. In 2000, another Schgeer-sponsored law increased the price for solar energy and launched the installation of 100,000 solar panels on homes and businesses. In June he orchestrated a law eliminating taxes on bio-fuels, such as gasoline substitutes made from plants. That's a significant benefit in a nation where gasoline costs $4 a gallon. Meanwhile, it was Germany's example that inspired the European Parliament to mandate a doubling of the use of renewable power across the continent.

Schgeer, who has preached his gospel in 100 countries, does not believe the world can afford to wait for the market alone to make wind and solar power compete with fossil fuels. Renewable energy, he says, is necessary for the assurance of life on earth said Margot Roosevelt.

Chapter 7: Factor Farming and Waste

The small farmer living an idyllic pastoral life has become virtually nonexistent, being displaced by large conglomerates that conduct their farm operations like any other big efficient business. There are no clucking hens walking around singing in the barn yard and no hogs and pigs wallowing in mud and drying in the sun. These industrialized farming techniques are commonly referred to as "intensive farming" and "factory farming," Although the phrases are often used interchangeably, they are quite different. Intensive farming involves increasing productivity through better management and breeding techniques, without significantly changing the pattern of life the animals lead. Factory farming alters the pattern of the animal's life and results in undue physical pain and mental suffering.

Factory farming is characterized by overcrowding, restricted movement, and unnatural diets. The procedures utilized intensive farming procedures in such a way that results in severe suffering for the farm animal. The poultry industry uses factory farming techniques in nearly every phase of poultry production that illustrates the abuses. Factory farming occurs in the raising of all farm animals, hogs, calves, cattle, and chickens. This section presents several examples of factory farming techniques and the cruelties they impose on farm animals. It also comments on the harmful pollution that gets into our water and the atmosphere because of factory farming.

Chickens

Two types of chickens are raised in factory farming systems, laying hens, which are grown for egg production, but which are also subsequently used in soups, stews, and pot pies. Broiler chickens are raised solely for consumption. The methods share some of the same features. Both broiler production and laying hen production begin at the primary breeder, a laboratory that develops the genetically different strains of chicken. Broiler birds are bred to gain weight rapidly and have good body conformation while laying hens are bred to lay thick shelled eggs with thick yellow molds. The multiplier hatchery, the second stage in the production of both kinds of birds, consists of large sheds containing ten to fourteen thousand chickens called breeders. The breeders, a third strain of chicken, lay fertile eggs which hatch into laying or broiler chickens and, unlike the two other strains of chicken, are not retailed to the consumer.

Laying Hens

If the hatchery breeds laying hens, workers separate and eliminate the male chicks at hatching because they cannot lay eggs and thus lose their economic usefulness. The male chicks are either dumped alive into trash bags and left to suffocate or are drowned. The dead chicks are then used in the manufacture of various animal feeds.

In the breeding farms, they are kept in darkness or near darkness until they are ready to lay eggs, normally a period of approximately twenty weeks in darkness which is about 5 months. When the birds begin to lay eggs the lights are turned on conditioning the hens to lay eggs whenever the lights are on. Each week the lights are left on for progressively longer periods of time until after forty weeks, they are on seventeen hours per day. This lengthening of the birds day increases egg-laying productivity and generates greater profits. Such a regimen takes its toll. While chickens raised under natural conditions can lay eggs for as long as twenty years, laying hens subject to these artificial conditions exhaust their laying capacity after only one or two years.

Laying hens live under even greater conditions of stress than broilers. The battery cages are more crowded than broiler sheds, ranging in size from one to four and a half cubic feet, and in occupancy from four chickens in the smaller cages to nine chickens in the larger ones. Cage life creates great physical discomfort for the animal. Cage floors are made of wire and while this facilitates cleaning, it runs counter to the instinctual need of hens to scratch dirt. Since this instinctual scratching wear down their toe nails, chickens confined in wire cages often grow toe nails so long that they become entangled with the cage and are literally grown fast to the cage.

Cage life creates conditions known as cage layer fatigue sympathized by brittle bones, inability to stand, and pale with washed out appearance. In addition, breast blisters, foot pad lesions, feather follicle infection, and feather loss all are commonly suffered by caged hens. Studies indicate a ten to fifteen percent death rate among chickens raised under these conditions. Some of these deaths result from fighting and cannibalism, and growers subjecting laying hens to de-beaking, although the longer life spanned of laying hens means that the operations must often be performed twice.

Broilers

Raising tens of thousands of broiler chickens in a confined space exposes them to various threats of contamination. Infected fecal material in liter can contaminate the flock with Campylobacter and Salmonella.

The third phase of production entails the use of breeding farms, where the chicks are kept until maturity. The grower uses artificial lighting for unnaturally long or short periods of time in order to produce certain behavior in the birds. In broiler breeding farms, bright lights encourage the chicks to start feeding, while dimmed lights reduce the stress cause by over crowding as the birds mature and increase in size.

After the breeding farm, broiling chickens are shipped to growing farms and placed in broiler sheds, while the laying hens are sent to laying farms and placed in better cages. These birds that spend time in these surroundings and under these conditions, encounter most of their suffering.

From ten to fifteen thousand broilers live on the floor of each broiler shed. Life in the shed is completely automated in order to enhance the bird's growth. No natural light enters the shed. Instead, an automatic mechanism adjusts artificial light depending on the need, either brightening the light to induce the birds to eat, or dimming it to reduce the effects of overcrowding. Hoppers suspended from the roof of the shed automatically dispense food and water.

By the time the chicken is ready for slaughter, the overcrowding in the shed is so sever that only half a square foot of floor space remains for each bird, creating a high level of stress that manifests itself in outbreaks of fighting and cannibalism. If a sudden change occurs in the shed because of a variation in the lighting or the entry of human intruders, the chickens panic and rush to one corner of the shed, piling on top of each other and suffocating those on the bottom.

Chopping and Burning off Their Beaks

The most logical solution to cannibalism would be more room from the crowded conditions. But this is less economical than the present growing techniques and therefore unacceptable to agribusiness. The economic starved growers use solutions that create even more suffering for the poultry. As a way of controlling the cannibalism that results from overcrowding, chicken farmers routinely include drugs in the chicken feed. In addition, they remove the chicken main defense weapon, the beaks. De-beaching is an extremely painful process, accomplished either with a guillotine-type device that chops off the beak, or with a hot knife machine that burns it off.

Growers rationalize the use of battery cages by citing their alleged necessity for maintaining a reasonable price for eggs. More humanely designed cages do exist and they are just as efficient as battery cages. These alternative cages, about one cubic meter in size, have two tiers. The lower level contains nest boxes in which the birds lay their eggs and the second level contains perches, for food and water. A six-month study that compared the new cage with the traditional battery cage found not only that egg production remained equivalent, but also that the birds in the new experimental cages did less pecking, pushing, and other problems cause by stress. Unfortunately, without the impetus of legal compulsion, farmers have displayed no willingness to use these cages.

Loading Chickens for Market

Chicken's suffering does not terminate at the farm. Broilers are loaded into trucks in a variety of inhumane ways. Some farmers catch birds by hand, others use bulldozers-like devices to force the birds into the trucks, and in Europe, a "vacuum" machine has been developed which actually sucks the birds through a large hose into waiting crates. Laying hens receive somewhat better treatment. Growers transport them to market in their cages. The slaughter of both laying hens and broilers, are unnecessarily brutal. The birds are unloaded from the trucks and hung upside down on a conveyor belt as they await slaughter.

At the slaughter house chickens whose intestines are accidentally pierced by gutting hooks are removed from the assembly line and processed elsewhere, if they are seen. But contaminated mechanical fingers used for plucking can press bacteria into the skin. Pathogen may also spread in the chilled "water baths" in which carcasses are plunged.

Usually packaged at the processing plant, chicken may spread contamination if juices leak in shipment. If a chicken is improperly refrigerated at the store, bacteria will multiply, and studies show that a majority of store chickens contain pathogens, a disease.

If the chicken is infected by feces in food, the food will carry Salmonella bacteria. Chicken and eggs are especially high-risk carriers, but an emerging strain shows high rates of drug resistance.

Factory farm operations have become widespread in the poultry industry with ninety-eight percent of all broilers raised by such systems. Yet these factory farm operations by no means are exclusively to the poultry industry. Agribusinesses are utilizing similar confinement systems in the production of larger farm animals such as hogs, calves, and cattle.

Texas Cattle Farms

Rainy Creek that runs into the Middle Basque River about six miles away is the North Basque River and two counties over is Earth County. It is home to at least 250 factory dairy farms called CAFOS, for Confined-Animal Feeding Operations. The CAFOS milk as many as 2,000 cows a day and the county has about 110,000 dairy cows that produce an estimated "1.8 million tons" of cow manure a year, 34,615 tons a week, 4,495 tons per day. The manure has gotten into the North Basque River and its tributary streams which feed into Lake Waco, the drinking water source for the city of Waco. The local water in Earth County shows increasing levels of nitrates, ammonia, and fecal coli form bacteria.

A farmer hired an independent water-monitoring firm and learned fecal coli form counts in his creek were running from 50,000 units per 100 milliliters to millions and even billions of units. The maximum is supposed to be 200. The increased phosphorus downriver threatens the water quality for the whole area. Jony Young, editor of the Waco Tribune-Herald says, the water is not fit for carp which is a fish that eats garbage found in the water. He also says there have been so many taste and odor events, the euphemism for bad days that the town should fly a green flag whenever the water tastes like water.

Texas, we are proud to report, ranks No. 1 in the country for animal-waste production, creating an estimated 280 billion lbs of manure annually, which is twice the amount of manure California produces, the No. 2 state. That comes to 40 lb. of manure per Texan per day. The state is covered in negative glory.

Many of the CAFOS are owned by people from the Netherlands, who came in droves for the cheap land, high milk prices, and lack of regulation. One result is growing animosity in the region against the Dutch, which, if you did not know about the cow manure, might strike you as an odd development. The industrialization of dairies, a national phenomenon, mirrors changes in the poultry and pig industries. In the Panhandle, the problem is pig manure, with exactly the same results, except there they are mad at the Japanese, who owns many of the corporate farms. The pig manure gets into the lakes and is starting to affect the Ogallala Aquifer. In East Texas it is chicken factories. For Texas there are dairy farms and dairy farms.

In December 1998, the Natural Resources Defense Council issued a report detailing how cow manure in central Texas is poisoning drinking water supplies by fouling underground aquifers as well as rivers, lakes, and streams. It has gotten into the North Bosque and its tributary streams which feed into Lake Waco, the drinking-water source for the city of Waco. The local water in Earth County shows increasing levels of nitrates, ammonia, and fecal cloakroom bacteria.

How Are Cattle Infected With Pathogens?

The high cost of grain makes it more profitable to graze cattle on open fields during their first two years of life. Nevertheless, beef production does not lack its share of abuse.

Conditions in these concentrated animal cities favor contamination and the spread of disease. Cattlemen have begun to adopt many of the procedures utilized in the production of other animals. The most notable changes in raising cattle have occurred in the feed lot where cattle are placed for fattening for their final six months before slaughter. Although feed lots in general still consist of open, outdoor lots, an unfortunate trend has begun toward total confinement buildings. The space allocations are as inadequate in indoor lots as in other confinement systems. Their floors are often inches deep in a soupy manure mixture which densely cakes the animals coats. They are jammed together standing on black stuff which is all feces. When they reach the slaughter house, they are covered with feces and still crowded together.

When beef is processed into ground beef, the chances of contamination rise significantly. Processing meat contaminated by one animal can spread the pathogen to the hamburger that passes through the machinery in one day. The USDA inspects every carcass in every meat and poultry processing plant, but without checking for microbial pathogens. Meat inspectors still rely on sight, touch, and smell to spot disease says Mike Taylor, former administrator of USDA's Food Safety and Inspection Service. However, bacteria that make people sick cannot be found that way.

Scientists from the USDA visited four large slaughter houses in the Midwest to test beef cattle for E. coli contamination in the summer of 1999. They found 28 percent of the cattle entering slaughter houses were infected and 43 percent of the skinned carcasses were contaminated, but by the end of the processing, only 2 percent of that tested meat was tainted suggesting measures had been taken by meat processors to reduce the contamination.

Many meat processing plants use procedures that include chemical baths, rinses, and sprays, bathing carcasses in steam, or irradiating processed meat to kill microbes. But it could be still made lower by farmers reducing infection in their livestock. The problems arise from manure. We are extremely careful with human feces, but cattle feces works it way into streams and groundwater, which we use to irrigate and wash our produce. Manure is also used as fertilizer. If it contains E. coli and Salmonella, we are re-circulating these pathogens through our environment.

Pathogen is a virus or a disease and when farmers shift the diets of beef cattle from hay to grain in order to boost growth rates and reduce costs does this cause pathogens? When ruminants are fed fiber-deficient rations, wrote USDA's James B. Russell and Jennifer Rychlik of Cornell University, microbial ecology is altered, and the animal becomes more susceptible to metabolic disorders and, in some cases, infectious diseases.

New technologies have encouraged the feeding of a wider range of materials to cattle, including "wastes." Chickens in the United States eat a variety of feed, including fish meal from Asia. Cattle eat such agricultural by-products as peanut hulls, almond shells, wastes from bakeries, poultry manure, and it is shipped all over the world. This is creating new niches and opportunities for food borne pathogens to enter the food supply and spread.

Great Britain had evidence of the dangers of using animal by-products in livestock feed that surfaced in the out break of mad cow disease. The rapid spread of the illness, which likely resulted from feeding cattle meat and binomial from animals that already had the disease, was linked with more than a hundred cases of deadly Creutzfeldt-Jakob brain disease in humans who had consumed the infected meat. Since the outbreak among cattle in Great Britain in 1986, BSE has been found in animals in several European countries and Japan.

From 1997, the FDA banned the use of rendered "remains of dead cattle and sheep in feed" for United States ruminants, and there is no sign of BSE in the United States. But regulations still allow the use of animal blood and blood products as well as pig and horse protein. They also allow poultry to be used in cattle feed and cattle to be used in poultry feed. Is this an effective recycling of animal protein or a breach in a basic ecological relationship, with serious consequences for our food supply?

Also farmers have been adding antibiotics to animal feed for more than half a century, after it was discovered that the drugs were effective in accelerating the growth of animals. Now by some estimates the volume of antibiotics used in animal feed equals or exceeds that used in human medicine.

The use of antibiotics as food supplements for farm animals is a serious threat to human health, says Alicia Anderson, an epidemiologist for the National Antimicrobial Resistance Monitoring System. She and others believe that use of the drugs in healthy animals is playing a role in changing the very nature of food borne bacteria, creating strains that are resistant to antibiotics used in human medicine.

Since the early 1990s infections with the super bug DT104 and other food borne antibiotics resistance bacteria have turned up in several countries. A report published in 2001 after scientists at the University of Maryland and the FDA sampled ground beef, turkey, chicken, and pork from supermarkets in Washington, D.C., revealed that a fifth of the samples contained Salmonella, and 84 percent of these organisms were resistant to at least one kind of antibiotic. Some were resistant to as many as twelve.

Cattle and steers must all endure other abuses such as hot iron branding and castration. The castration process is as follows. The procedure is to pin the animal down, take a knife and slit the scrotum exposing the testicles. Then grab each testicle in turn and pull on it breaking the cord that attaches it. On older animals it may be necessary to cut the cord. Cattlemen rationalize this procedure by stating that steers gain weight more rapidly than bulls. Actually castration simply makes them put on more fat and nothing is said about the pain they endure. Male hogs are treated the same way.

Chickens, hogs, calves, and cattle are all forced to live under abnormal stressful conditions. The life of a farm animal does not resemble the lazy serene existence described in story books. It is a horror for which we as consumers must bear the ultimate responsibility.

Nebraska Farm Complaints

Even the gentlest southerly wind blowing through the Muddy creek Valley in Nebraska can make life unbearable for Bev Hurt. The breeze blowing past the Tumbleweed Cafe through the town square and past the miniature ghost town playground in Melham Park carries the stink of 85,000 cattle.

Rotten manure is the smell that greets you in the morning," said Hurt, a 55 year old studio photographer who hasn't opened the windows in her home in more than two years. Even with the house shut tight, the smell seeps in.

You do not want to go out and walk your dog, she said. You do not want to go outdoors at all. The putrid odor originates from the Adams Land & Cattle Company feedlot about five miles south of town, and it is especially pungent after the rains. The company operates another feedlot with about 8,000 cattle just east of town, leaving tiny Broken Bow with few directions the wind can blow without raising a stink.

In Nebraska, neighbors have found themselves in a no win battle against the stench from feedlots, meatpacking plants, and other businesses tied to Nebraska's $5 billion livestock industry. The offending odor is generally the same from the feedlots near tiny miniature in extreme west Nebraska to metropolitan Omaha's meat packing plants on the other end of the state.

The permeating smell is like the stench of rotten eggs. It is the smell of hydrogen sulfide, a gas produced by decaying organic matter such as at wastewater treatment plants, livestock operations, and meat processing plants. Exposure to the gas can cause headaches, fatigue, depression, and nausea. It can irritate eyes and cause nerve and respiratory problems.

Agribusiness is concerned solely with ensuring that its final market brings the most money, and is unsympathetic to the plight of farm animals. Thus, the life of a farm animal is not the same as seen on farms from the highways.

Communities across the state find themselves in the middle of fights among neighbors and businesses over smells from the livestock industry, but cattle trucks are prohibited from the stockyards "theater" parking lot in Omaha.

Feed lots are confinement facilities for beef cattle. Rather than grazing on pasture land, cattle are fed from central facilities. Feedlots are also dependent on chemical feed additives for weight gain in confinement and for temperament control of cattle in confined conditions. Feedlots have been criticized for increasing the risk of disease spread among herds, as well as ground water pollution from feedlot runoff.

Cattle feedlots are an economic possibility only because of a separate agricultural chemical topic, the use of antibiotics and synthetic hormones in livestock feed. Like pesticides, the recent history of chemical feed additives is one of immense growth. According to the Office of Technology Assessment, all poultry, seventy percent of beef cattle and veal calves, and ninety percent of swine raised in the United States receive some form of antibiotic feed additive. The United States feed additive industry is a $1.5 billion industry, with sales here and abroad. American Cyanamid produced and sold $120 million worth of antibiotics for animal use in the United States in 1981, and $265 million abroad.

There are two kinds of feed additives, antibiotics and synthetic hormones, and both are linked to a major trend in livestock production, keeping animals in confinement. Both antibiotics used in sub-therapeutic amounts, and hormones are used to stimulate weight gain. Some hormone injections are also used to alter behavior, preventing animals from harming each other in confinement.

The best-known synthetic hormone is DES, diethyl stilbestrol. It is used to stimulate weight gain in cattle, but this substance was found to be retained in meat consumed by humans. DES and another estrogen, ECP, estra idol cypionate have been associated with "ovarian and uterine cancer" in adults and childhood abnormalities. The antibiotics in the beef eaten builds up in the human and the antibiotics taken by them prove to be "ineffective."

Feedlots make many evenings ripe in Wisner, Fairbury, and Dorchester. Meatpacking plants raise a stink in Gering, Schuyler, and Madison. The occasional whiff of burnt barbecue can be expected to filter in from Omaha's suburbs, but moviegoers leaving the Stockyards Theater on the city's Southside soon realize why they paid only $1 for a seat. The theater near what used to be the world's largest stockyard is next to several meatpacking plants.

It is not just the smell of money, anymore," said J. D. Alexander, president of the Nebraska Cattlemen. It is reality, and something we have to deal with in our profession to be successful. The odors got so bad in Grand Island that the city set up a hot line that residents could use to call.

The ConAgra Beef Company, in Greeley, Colorado in 2000 recalled 19 million pounds of beef. In the second largest meat recall in United States history, a Colorado company asked Americans nationwide to check their refrigerators, stores, and backyard grills and destroy 19 million pounds of hamburger meat because of E. coli concerns.

Seventeen people in Colorado became sick after eating beef, and six other cases of E. coli caused illnesses have been reported in California, South Dakota, Michigan, Washington, and Wyoming.

Fattening Cattle The Old Way

In the 1950s, cattle were grazed for a long period and fattened on grain at the end of their lives because it was the cheapest way for small farmers to produce them. Now the cheapest route to the packing plant is corn, as much and as early as possible. It is cheaper to fatten cattle quickly on large quantities of corn rather than let them slowly graze on their own.

Cows are ruminants and nature designed their systems to digest grass and hay, along with some grain in the grass seeds. When the cattle go to commercial feedlots, their digestive systems try to readjust slowly. On this diet, they can gain as much as three pounds a day, compared with about one pound on grass, and the time from birth to slaughter can be shortened, but a price is paid.

The acid produced by corn seriously affects the animal's digestive tracts and livers, especially if the animals are very young. To counteract that, antibiotics are usually administered. That adjustment, along with growth hormones, both are supposed to be stopped after a prescribed period before slaughter to remove the drugs from animal's systems. That means a steer that weights 450 to 600 pounds at weaning can shoot up to 1,200 pounds in seven to 10 months. The science on antibiotics and hormones, whether feeding them to cattle or pork affects those who eat the meat, is often debated. It is clear that their use has to do with economics.

Corn-fed beef is well-marbled, and as produced by commercial feeders, has a consistent taste. According to the Denver-based National Cattlemen's Beef Association, no one loses.

Farmers and ranchers go along with the commercial system because they have no alternative, and because they lost their initiative. Those who have the financial means to assume risk, usually the large corporations, set the rules and reap the biggest profits. No one has a clear vision of how to get out of this cycle.

Veal Calves-Non Profit

Animal welfare groups have been researching more humane alternatives for veal production without impairing the quality of the meat. One system would permit the calves to exercise, lay on straw, and be fed milk through teats on automatic milk machines. Unfortunately, farmers have not as yet demonstrated a likeness for this system.

They do not have outdoor pastures during good weather. Large agribusiness's implementing total confinement systems are rapidly displacing these small farms. Dairy operations use two kinds of confinement systems, free-stall holding barns and tie-stall holding barns.

In free stall holding barns the cows can move throughout a limited area within the barn, although they must walk on slippery, slatted floors. In tie-stall holding barns, a tether constrains the cow within a narrow stall. No reports have been published about the effect of these systems, although it is likely that the same problems associated with stress in other animals exist in the case of dairy cows.

They also have either a cement or slatted floor. No straw is provided for cushioning because straw contains iron which ingested, would cure the anemia necessary to create veal's pinkest tone. In Addition, the separation of the calf from its mother causes psychological harm. The cumulative effect creates great stress, making the calf susceptible to salmonella, diarrhea, and other infections. To prevent these diseases, farmers ordinarily add antibiotics and other drugs to the calf feed.

Veal production has earned interpretations as being the most morally repugnant factory farm operation, comparable only with barbarities like the force-feeding of geese through a fuller that produces deformed livers. The origins of the veal industry began in dairy farming, where a farmer would slaughter unwanted bull calves, prior to weaning, for use by his own family. Veal's pink color and extremely tender quality made its sale to consumers quite attractive, although each young calf had insufficient meat to make such efforts profitable. However, the demand for veal spurred the development of a system that fattened the calf but simultaneously maintained its premature condition by denying the calf any exercise, maintained it in semi-darkness, and provided a diet designed to make it anemic.

In modern husbandry systems, the calf spends its entire life in a tiny stall in which it can neither turn around nor lie down normally. Instead, the calf must lie in an uncomfortable haunch position.

Hog Factory

In recent years, hog farming has developed into a total confinement systems, which range from a two phase breeding-growing system to more elaborate multiphase operations. In the breeding phase, the pregnant sow generally remains in an individual stall until about a week before she is ready to give birth. She is then moved to a furrowing stall where she gives birth and nurse her piglets for about a week. In order to restrict the sow's movement, the breeding and furrowing pens are kept quite small. The pens permit the animal to stand up and lie down, but prohibit the sow from turning around. Many farms have begun to use iron frame devices known as "iron maidens" which prevent the sow from moving at all. In effect, she becomes a living reproduction machine.

The United States Department of Agriculture supports the used of furrowing pens as a safety device, reasoning that the young piglet would have little chance of surviving if its 500 pound mother accidentally rolled over it. However, the department fails to recognize that the sow's confinement to such a small area is the cause of such a danger. Moreover, the piglet which the department seeks to protect suffers abuse just like its mother. In a day or two of birth, "the young piglet has its ears notched, its teeth clipped, its tail docked and, if male, is castrated as well." Tail docked is to cutoff part of the tail.

In a two-phase system the farmer transfers the piglets when they are five or six weeks old to a finishing pen where they are fattened for the next thirteen to fifteen weeks before being sent to market. In a multiphase system, the pig first goes through a nursery phase before being sent to the finishing pen.

Although they vary in size, finishing pens allow no more than six square feet per pig which the pen will be three feet by two feet not giving much room to stand or lay down. While some pens are outdoors and have cement floors, the more modern ones are indoors and have either slatted floors of sloping concrete floors to facilitate cleaning. Such floors, besides being uncomfortable, damage the hogs feet and legs since they are unsuited for such hard surfaces. Hogs raised in dirt pens, on the other hand, show only minor damage.

Although the overcrowding of hogs does not reach as severe a level as that of chicks, it is high enough to produce stressful behavior.

The hogs bite each other's tails, a phenomenon the farmers have tried to control with chemical food additives and tail-docking, cutting tail off. Stress also manifests itself as a physical condition known as the "porcine stress syndrome" noticed by such symptoms as rigidity, blitch skin, panting, anxiety, and often sudden death.

Hogs possess high degrees of intelligence, remarkable similar to those of dogs. If farmers raised dogs in the same manner in which they raised hogs, prosecutions for cruelty to animals would surely result. Yet no such prosecutions are imminent for pork producers. On the contrary, the government endorses current growing techniques and actively supports research to develop systems which would offer the potential for greatly increased animal capacity in front line facilities.

Hog Factories in Missouri and Kansas

There is a threat to our drinking water by manure run off from small farms and giant farms. The manure is generated by cattle, poultry, and large hog farms and the trend toward industrial scale farming has created an enormous increase in the concentration and quantity of manure that is generated at a single site. As the number of hogs raised on a single farm has gone up, so has the amount of manure contained at one site. In some instances, manure has contaminated ground water either by a leak from a storage tank, lagoon, or when rain falls have swept the manure off farm fields where it was spread as fertilizer.

The centers for Disease Control and Prevention of three women in LaGrange County, Indiana, had six miscarriages from 1991 to 1993. The miscarriages were traced to the nitrates in the water from manure produced at a nearby hog farm and nothing was done about it.

A hog farm in Missouri near St. Louis discharges more waste in the Missouri river in a year than "all of St. Louis."

The hog farms are affecting Kansas, Texas, and Oklahoma in the Middle West that is ideal for the actions this company takes. Seaboard Corporation is a giant of agribusiness with headquarters in Merriam, Kansas and was controlled out of Chestnut Hill, MA not far from our home.

For most of this century, Wilson Foods operated a pork plant and was the town's largest employer. In the early 1980s it had to cut workers' average annual pay from $22,200 to $16,200 and eventually sold the plant to Farmstead Foods. Farmstead lost its biggest customer, Wilson, and when the workers were receiving their last unemployment checks, Seaboard appeared. Seaboard announced it would restart the closed pork-processing plant that had once been Albert Lea, Minnesota's largest employer only if the city offered financial help. The city responded by giving Seaboard a $2.9 million low interest loan and special deals on its sewer bill, grading, and paving parking lots for employees.

After a lot of promises, the company received sweetheart deals with the city and the Chamber of Commerce erected a billboard saying, "35,000 Friendly People Welcome Seaboard Corporation." Since hog killing created serious pollution problems, Albert Leas earlier had added $3.4 million to build a waste water treatment plant devoted to servicing the pig factory. The Federal Government contributed $25.5 million, while the state of Minnesota gave $5.1 million. The total cost of the sewage plant was $35 million free to Seaboard and to add to that the city built new roads and water lines to the plant, built a parking lot, and came up with another $1 million to help erect a hog slaughtering building with a total cost free to Seaboard. This is the way big businesses decide where they will bill a plant. What city gives them the most money.

Seaboard failed to invest in upgrading its sewage pre-treatment facility and its waste began to overwhelm the city's municipal treatment plant. Normally the city placed its treated sludge on soybean cropland, but by the second summer, city officials were in search of more land. Sparks said, "We had so much sludge accumulation that we had to go out in the middle of the summer and buy a crop for $36,000 and plow it under and place the sludge on top of the ground because our storage capacity was exceeded."

In August of 1992, Seaboard announced it would employ as many as 1,300 workers at its new pork-production facility. The company would slaughter 4 million pigs a year and Governor David Walter's declared the plant "a huge and much deserved economic boost to the entire panhandle area and the state." That is 10,959 pigs per day or 457 pigs per hour based on a 24 hour day. At the same time in August 1992, the Seaboard's president was reassuring newspapers that the Albert Lea plant would remain open.

Seventeen months later in January, Seaboard announced that it would close its hog-slaughtering operations and lay off about 600 employees. The number of employees dropped to about 200 and Seaboard sold the business making a profit. It moved its hog-slaughtering operations to another small town Guymon, Oklahoma 800 miles away with a larger corporate welfare package. It was named Albert Lea and the work force was unionized and the wages were $19,100 a year, $3,000 below the 1983 wages.

Oklahoma provided $21 million for Seaboard to come to Guymon, but Guymon could not supply the work force required by Seaboard. Since the turnover rate in all processing plants runs close to 100% a year, the company needed workers by the thousands. The slaughter house was one of the world's largest and would kill an average of eight hogs a minute 24 hours a day 364 days a year that adds up to nearly 4 million annually.

It hired immigrant workers, Laotian and Vietnamese, most from Mexico, Guatemala, and Honduras, other central and South American countries, and illegal immigrants. Guymon is located 320 miles east of Santa Fe. MN, 335 miles west of Tulsa, 125 miles north of Amarillo, Texas and 500 miles from the Mexican border. The nearest bus stops were in Liberal, Kansas which is 40 miles to the north of Stratford, Texas.

Operating Hog Factories

The company was large and operated huge hog farms in five southwestern counties in Kansas that accounted for more than one-quarter of the state's 1.5 million pig population. The pigs were raised in Kansas until ready for slaughter and trucked to the processing point in Guymon, Oklahoma.

Usually farms are thought of as the ones seen when traveling the highways, but hog farms are pig factories. The barns are long and houses about 1,000 animals jammed next to one another eating constantly until they grow from about 55 pounds to 250 pounds. The floors are slated so wastes drop into a trough below that is flushed periodically into a nearby cesspool. From 1990 to 1998, the Oklahoma hog population increased 761 percent from 230,000 to 1.98 million with Seaboard accounting for about 80 percent of that number.

The strong smell from Seaboard's 40,000 hogs closely confined in 44 metal building used exhaust fans twenty four hours a day that pumped out tones of pungent ammonia mixed with tons of grain dust and fecal matter, scented with the noxious odor of hydrogen sulfide, a poisonous gas produced by decaying manure that smells like rotten eggs, all combined with another blend of aromas wafting from five cesspools each 25 feet deep and the size of a football field. They can be considered open-air sewage ponds and to compound the problems, 75 feet below lies the Ogallala aquifer which provides "drinking and irrigation water for a large part of the county."

Dead Hogs Laying Around Rotten

An additional foul odor was generated form dead pigs and hogs just laying around. The written law said the carcasses are supposed to be deposited in dumpsters with the "lids tightly closed and the contents disposed of daily." Seaboard fell behind in disposing of thousands of hogs dying before their time each year and where did they dispose of them? The overflow of dead hogs were stacked near the dumpsters, piled besides the large barns, besides roads, and sometime they were not disposed of before their flesh rotted away and the flies and other insects got fat while spreading diseases.

The number of death for hogs was high. The dead-pig truck or trucks pick up hogs once a day if on schedule otherwise there were usually hundreds of dead and decaying hogs. The Oklahoma agriculture department fined Seaboard $157,500 in December 1997 for improper disposal. The company appealed and paid the state only $88,200.

Julia Howell and her husband Bob lived on a farm near Hooker that is midway between Guymon, Oklahoma and Liberal, Kansas. Four generations of Howells have grown wheat and raised families, but now Julia Howell which is 69 talked about her 40,000 "neighbors," hogs, and explains why she seals the farmhouse windows, stuffs pillows into the fireplace and seldom ventures outdoors without a face mask.

For two years Julia has recorded in a diary, life with the blended smells from rottening hogs, cesspools, and the breezes from hog barns: "Monday, July 1, 1996: 80 degrees F, calm, Tried to sit outside a while. Impossible without a mask, what a life! Monday, July 1, 1996, had a storm at 70 degrees F. It rained toxic fumes at 7:30 pm. Horrible during rain! Wednesday, July 24, 1996, calm, 80 degrees F. 9:30 pm. It would take two masks tonight."

We celebrated our 50th anniversary here this year, but when the hog fumes came rolling in, we could not plan on anything. I have not had people in for dinner for two years because I probably would have to meet them on the driveway with a mask for them to get to the house. Seaboard tells them your home is worth more than you ever dreamed because of us coming in next to you. Julia says our kids could not sell the property if they needed the money to bury us. It just devaluated to nothing as far as the market's concerned.

Vancy Elliot and her husband lives about 3 miles from Guymon and their land abuts Seaboard's hog farm. They have flies in the winter and put fly traps out in the summer. Rats and mice are a real problem because so many pigs die.

Two Separate Hog Farms

In northeastern Iowa there is a hog farmer named Gary Lynch with steel sheds that hold 1,000 animals each with a total of 100,000 or more hogs spread over six counties. The state has 15 million pigs on their farms and Lynch is proud of his business and to show it he has Porkie on his car license plate. Tom Frantzen 30 miles away, took over a family farm in New Hampton in the mid 1970s raising organic pigs, unfettered, variably mottled and plump that sleep on the ground or beds of straw in tent shelters. Both farmers transform 650 pounds of feed in six months to 260 pounds of pig that includes 1,200 pounds of waste that is left behind.

The USDA Organic labeling rules require farmers to give their animals access to the outdoors, but the pigs are not required to take advantage of it, and antibiotics and artificial feed additives are prohibited. The pigs are fed organic food and are fed on Frantzen's own feed that he grows. Since organic feed costs 50 to 75 percent more than conventional feed, his crops helps in the feeding. When the pigs get sick, he treats them with natural remedies like oak bark and walnut husks.

Lynch raised 250,000 hogs which is about the total number that all organic farmers combined raised. Frantzen raised 1,050 organic hogs. He hopes the new labeling rules for organic hogs will raise the figure to 1.8 percent which would be on par with organic milk.

There were 97 million hogs raised in America last year, and George Siemon's of Organic Valley, the farmer-owned cooperative that markets Frantzen's hogs thinks that less than 1 percent of the total herds were organic.

Tom Frantzen's Operation	Gary Lynch's Corporate Farming
1. Open space for laying on ground, grass, mud, or in sunshine.	1. Enclosed metal building, no natural light,
2. Natural insemination.	2. Artificial insemination, cheaper.
3. Bred by Berkshires, darker juicier meat.	3. Bred by Landrace, Duroe, Hampshire mix and other white meat.
4. Delivery, 7 to 9 piglets on bed of straw in plywood shelter & natural pasture.	4. Delivery, a litter of 11 in a pen 5 by 7 ft. designed to prevent sow roll over on pigs.
5. Suckle six to eight weeks living in plywood shelters.	5. Suckle no longer than 3 weeks.
6. Exposed to freedom, sunlight, grass, & mud.	6. Remain in their pens.
7. No antibodies in sow's milk. Gain weight normally faster.	7. Antibodies in sow's milk, gains weight.
8. Leave tails, showing good upbringing.	8. Clip their tails that show overstressed.
9. After weaning, pigs houses have floors with crushed limestone, open at both ends and holds 140 pigs in a loose huddle.	9. Sheds divided into pens 19 ft by 10 with 24 pigs each.
10. Puts straw on floor to soak up dung and urine.	10. Slat floor above a 500,000-gallon waste pit of concrete and pumped twice a year.

Clipping tails was a common precaution against pigs' tendency to bite one another on that tender part of the body. It is viewed as a cry for help from overstressed overcrowded pigs and is used to define their upbringing.

Pig manure is foul-smelling and loaded with E. coli with potent concentration of nitrates and phosphates. It is used as a fertilizer on land, but when it washes into streams, lakes, or rivers its bloom of algae is deadly to fish. Lynch twice a year applies its waste directly to his fields injecting the waste through a nozzle six inches below the ground surface. Some farmers spread the manure on the ground, spray it or impound it in giant lagoons that hold up to 20 million gallons that will leak or overflow in storms with the overflow going into lakes or streams.

The owner of the organic pigs puts down straw to soak up the dung and urine, and when the straw is saturated he puts down more. He believes that it is the key to controlling notoriously pungent like smell. When he sells a load of pigs, he spreads the straw on his fields, along with crushed eggshells, turning it in a rough approximation of composting with not a strong smell.

All pig farmers struggle to break even in the face of dismal returns that have reduced the number of pig farms in the United States by 90 percent since 1969. The profit is poor. After six months the profit on a 260 pound hog comes to all of $10.

Fish Farming

The leaper and chrome-colored fish that fights its way up northern rivers, jumping rapids and waterfalls on its spawning run have declined for decades and today the North Atlantic is dominated by a new kind of salmon, the farmed salmon. They outnumber wild salmon by 300 or 400 to one. In Norway, the world's largest population of wild Atlantic salmon, a single fish farm produces as many salmon a year as the estimated 600,000 wild salmon that migrate up the country's 650 salmon rivers to spawn.

Rivers in the United States were once home to an estimated half a million Atlantic salmon, but now the wild salmon are down to a few hundred and are continuing to decline in degradation of rivers, acid rain, netting of juvenile salmon by herring and mackerel trawlers, and shrinking ocean habitat.

Scientists blame it all on the industrial revolution with pollution, and dams that started in the 20th century. Before that age, it is estimated that at least ten million salmon returned annual from the sea too an arc of rivers that ran from the Hudson in New York, up through New England and eastern Canada, across to Ireland and the British Isles, over to Scandinavia and the Baltic, up to northern Russia, and down the Atlantic coast of Europe to Portugal.

Fish farming is becoming a bigger enterprise than beef or pork farming and it is destroying land along coasts and hastening the disappearing of the wild fish. Inland shrimp farms in Mexico look like more than 42 square miles of the Sonoran with patches of blue pools of shrimp. The shrimp farms are right next to each other all down the coast. There are ten salmon cages at a farm near Vancouver 100-feet square, eighty feet deep, and holds 100,000 baby fish. The distribution of food pellets is from spinning cones suspended above floats which are computer-controlled.

In the year of 2000, Mexico farmed 64 million pounds or 32,000 tones of shrimp that made their way to the United States. Most of the shrimp Americans eat come from abroad, and chances are excellent that they were farmed in Asia, Central or South America, or Mexico.

We also eat salmon raised on ranches that float in the seas off the coast of Norway, Chile, Maine, and the Pacific Northwest. The United States Food and Drug Administration say one-third of the seafood we eat is not wild. It really comes from aquaculture which is a $52 billion-a-year global enterprise involving more than 220 species of fish and shellfish that is growing faster than any other food industry.

Instead of helping save wild fish, aquaculture may be hastening the disappearance of them. Stanford University economist Rosamond Naylor discovered that a number of farmers in South East Asia have begun to raise shrimp because it could be at least 10 times more lucrative than growing rice. She thinks educating farmers and helping them be more efficient could help reduce the use of antibiotics. In the past 10 years, antibiotic use has dropped between 70 and 90 percent. But the United States Institute for Agriculture and Trade Policy estimates that between 204,000 and 433,000 pounds are still used every year in the United States alone. It is less than that fed to livestock, but still plenty to allow mobile waterborne germs to learn to mutate against the effects of antibiotics. To eliminate the problem, it is recommended that government prohibit the inclusion of antibiotics in any manufactured feed.

A bulldozer digs into a field to create clusters of five acre ponds on plots of land that range from 15 to 300 acres. A pipe is put at one end of the pond and pulls clean ocean water in, and a pipe at another end pushes water out laden with shrimp feces, excess shrimp food, and sometimes antibiotic into the ocean.

Most of the farms sits on an estuary where salt water and freshwater mix with an island nearby healthy and thick with Dolphins, irrigate birds, mangroves, Cormorants, Pelicans, and Blue Heron but it is killing the fish. Clams, scallops and mussels filter water for food. A three-inch oyster can filter about six gallons a day.

The intake pipes can pull in shrimp larvae, robbing fishermen of future catches. Excrement from shrimp and other cultivated species, including salmon, carp, tilapia, and catfish can sully water adjacent to farms, driving away wild fish and other sea creatures. Feces rich in nitrogen trigger the growth of algae, which can clutter and then choke the oxygen in bodies of water. The effluent can also release pests and diseases such as sea lice and viruses that thrive when shrimp or fish are packed together and can infect wild fish.

Hundred of millions of salmon in cages have fouled the sea around their pens, spread diseases and sea lice to wild salmon, and a large numbers of fish escaped. Tightly packed pens prove ideal breeding ground for sea lice, naturally occurring parasites that have devastated some salmon and sea trout population in Europe. Scotland and Norway studies indicate that sea lice out breaks at fish farms can have devastating effects on wild salmon and sea trout which are related species. Wild fish can pick up lice as they swim past infested pens and the deadly disease can be picked up by wild salmon from escaped farmed salmon.

In one Scotland farm about 50 million farmed Atlantic salmon swim around and round in pens as they are fed pellets to speed their growth, pigments to mimic the pink of wild salmon flesh, and pesticides to kill the lice that go hand-in-hand with an industrial feedlot. The pellets are shot through pneumatic pipes and hitting the water they sound like a hard rain as the fish chase the pellets. One 8 acre farm fish devoured 531 of the 976 pounds of pellets that was allotted for that day. It takes four pounds of fish rendered into food pellets to produce a pound of farmed salmon.

Fish that escape when seals chew through pens in search of an easy meal, when storms demolish cages, or fish spilled during handling can spread lice that kill fish by grazing on their flesh. Since over 300 million farm salmon are being sold every year, the magnitude of the escape problem is enormous when transferring the fish. In Scotland nearly 300,000 farmed fish escaped last year and 10 to 35 percent of the salmon on the spawning grounds of Norwegian rivers are farmed salmon meaning the pollution from the large population of farmed fish is reducing the wild salmon.

Shrimp and fish waste water may also contain antibiotics such as oxytetracycline and sulfadipethoxine, which is included in some fish food formulation says Stuart Levy, president of the no profit Alliance for the Prudent Use of Antibiotics. He and others worry that bacteria are developing resistance to these drugs. Ultimately, people will suffer if they encounter resistant strains of salmonella and escherichia coli which both are common causes of food poisoning.

It causes pollution and it leads to the destruction of coastal ecosystems because farmers clear out plants with bulldozers and alter the way water flows when they dump the displaced dirt and sand. Certain farmed species escape their ponds opening, consuming the resources of wild fish, and if inbreeding occurs, endangering the genetic strength of a wild species.

Biologist John Volpe of the University of Alberta estimated that in 1999 between 55,400 and 110,800 Atlantic salmon escaped from Canadian fish ranches which are made up of a series of 10 to 30 underwater 40 or 50 square-foot cages, each holding 20,000 salmon and anchored off the coast. They have been found in 17 Pacific Northwest Rivers so far, which means they are eating and breeding in the habitat of Pacific salmon and endangered steelhead trout. Biologists fear that the interlopers may ultimately push out the locals.

Salmon anemia forced farmers to kill 1.5 million salmon last year in Maine, and the disease has already been observed in wild Atlantic salmon. Carnivorous aquaculture species such as salmon, shrimp, eel, flounder, halibut, tuna, and sea bass are fed fish meal and fish oil, essentially ground or pressed anchovies, sardines, capelin, blue whiting, mackerel, Atlantic herring, and other small bony fish.

Each year roughly 66 billion pounds of these fish are used as feed, a growing percentage of which is going to aquaculture, 10 percent in 1988 and 35 percent in 1997. For some farmed marine species, as many as five pounds of wild fish are needed for one pound of growth. Farm fish could lead to the depletion in some regions of the small wild fish upon which much of the marine food chain depends.

John Hargraves, an Associate Professor in the Department of Wildlife and Fisheries at Mississippi State University thinks it is a very dangerous thing.

Chapter 8: No Help for Animal Treatment

Since these animals undergo excessive abuse, information is given to show what little positive feeling is given for the animals that help sustain our lives.

Legislators Approach too Animal Cruelty

Legislators approach the problem of animal protection through criminal anti-cruelty statues and regulatory statutes. Anti-cruelty statutes need to curb individual instances of cruelty to animals. Regulatory statutes address specific animal-related activities, such as hunting, selling, and trapping, and in some instances, seek to protect a species from extinction. An examination of federal and state legislation demonstrates that none of the abuses are occasioned by the factory farming of animals.

Regulatory statutes provide the main source of federal protection for animals. Some legislation seeks to conserve the existing stocks of a given species, while other legislation protects an animal after it has left the farm. No statute, knowingly, requitals farm animal treatment during the rearing process and no federal anti-cruelty statute exists to fill the void.

The Animal Welfare Act of 1970 provided for the humane markings and identification of animals, humane standards with respect to handling and housing, and investigations and inspections of animal conditions by the Secretary of Agriculture. The act expressly excludes farm animals from its coverage. The legislative history of the Act gives no explanation for the exclusion of farm animals, although it probably stems from the 1970 Act that was an amendment to the Laboratory Animal Welfare Act and thus farm animals were outside the scope of Congressional concern at the time and the lobbying must have been very heavy.

Although no federal legislation specifically protects farm animals, Congress has not altogether avoided the issue. A bill before the last few sessions of Congress proposed the establishment of a commission to study the treatment of animals in intensive farming. The commission would consist of eleven members, five members representing animal welfare and humane societies, five representing medial schools, veterinary medicine, animal husbandry, and one possessing administrative or judicial ability. The scope of the commission's investigators functions would not be limited solely to farm animals, but would also include laboratory research experimentation, domestic pet growth rates, and the effectiveness of existing laws.

The identification of economic alternatives is particularly important to animal welfare reform. While ample evidence exists concerning the current treatment of animals in factory farm systems, less data is available about alternative procedures. Certainly, a government report suggesting practical alternatives to factory farming would bolster the animal rights movement. Unfortunately, the proposed bill which would establish such a commission has never reached the floor of either house of Congress. Thus, at the present time, no federal law offers any protection for farm animals consequently, state statutes present the only possible source of protection.

Most state statutes define animals in very broad terms, similar to Florida's definition of animal as "every living dumb creature," that surely does not include humans. While in theory such broad definitions would seem to encompass intensively farmed animals, judges have used this vagueness as a basis for refusing enforcement and as a means of circumventing the statute. In one case involving a cruelty prosecution for cockfighting, the court held that birds were not within the purview of a state which prohibited the wounding of "an animal," noting that, although birds are, biologically spanking, animals, there was no clear legislative intent to include them within the statue. To hold otherwise, the court state, would "render the statute vague, indefinite and uncertain and therefore in violation of due process clause."

146

Definition of Cruelty

The second issue of statutory construction concerns the definition of cruelty. Cruelty is typically defined as "every act, omission, or neglect whereby unnecessary or unjustifiable pain, suffering, or death shall be caused or permitted." The crucial portion of this definition is the phrase "unnecessary or unjustifiable," since it makes the success of an anti-cruelty prosecution depend upon a showing that the contested factory farm practices are unnecessary.

The law provides little clarification as to what acts are unnecessary or unjustifiable. Those cases that do address the issue of the necessity of abusive farm techniques, although they are quite old and do not involve factory farming, do offer a few tentative interpretations. Some benefit must result to either the animal or to the community to justify painful farm procedures. Factory farming provides no benefits for the animal, therefore the issue turns on whether an economic sayings to the farmer is a benefit to the community.

Whether factory farm techniques are indeed unnecessary or unjustifiable has created a controversial and emotionally charged debate with economic and moral considerations. Proponents of factory farming assert that the price of food would be astronomical if factory farm systems were banned. The history of industrialization and mass production results in an almost automatic acceptance of this premise by the general public. Yet strong arguments support the conclusion that factory farming is unjustifiable and unnecessary.

First, the major costs of food production occur after the animal is slaughtered, with packaging, shipping, and marketing representing two-thirds of the retail cost. Savings in the growing of animals thus have a minimal impact on the actual cost to the consumer. Second, no definitive proof exists that the abusive factory farmer issuers any savings at all in the raising of animals. One study of egg production found that the stress produced by the overcrowded conditions

147

chickens are subject actually decreased the net income per bird. Third, efficient humane alternatives are available if factory farmers were willing to modify their systems; no such inclination is evident. Chicken farmers have failed to utilize the improved cage for laying hens. Finally, the moral question remains concerning the extent to which increased profits justify accompanying animal abuse.

Convincing a court that an abusive practice is unnecessary solves only a threshold issue. A second problem exists in those states that define cruelty as the unnecessary infliction of physical pain, suffering, or death. Factory farming not only entails instances of physical abuse, but also includes the infliction of mental abuse, such as the stress level in broiling sheds or the separation of the calf from its mother. If this issue ever arose in litigation, agribusiness interests would probably argue that the cruelty statutes are limited to physical abuse and that the law should not "recognize an animal's feelings." No reported cases have resulted in convictions under the cruelty statutes for causing emotional or mental deprivation. The question whether the mental abuse inflicted upon an animal comes within the definition of "cruel," remains open.

Excessive specifics or excessive generality do not present the only legislative deficiencies of anti-cruelty statutes. State statutes simply fail to proscribe specific factory farm abuses. All anti-cruelty statues fail to include overcrowding of farm animals as a forbidden activity. Although one case resulted in a successful prosecution for overcrowding under a broad anti-cruelty statute, the case involved dogs, not farm animals. A court would decline to extend such a precedent to the overcrowding of farm animals in the absence of specific statutory prescriptions, since the two situations are distinguishable on the basis of economic necessity. No economic necessity exists to keep dogs in an overcrowded condition, while agribusiness may argue present evidence of such a necessity in factory farming. Moreover, since the dogs in the above case died from the overcrowded conditions, there was a much stronger presumption of cruelty than would exist under merely stressful conditions. In reality, life under the stressful condition of factory farming produces more sever results for the animal than death.

Prosecutions based on the overcrowding of farm animals and absent of a statutory prohibition of overcrowding, seem unlikely.

No Exercise for Confined Animals

Some state statues seem to make certain factory farm practices illegal by requiring that exercise be provided for confined animals. Factory farming practices in the poultry, beef, veal, and pork industries thus could all be challenged under such provisions. No attempt to make use of these statutes in prosecuting factory farmers has ever been made.

Another example of inadequate enforcement of anti-cruelty laws concerns the neglect of injured animals under factory farming procedures. Since such farming procedures permit one person to maintain thousands of animals, injured or dead animals invariably go unnoticed for long periods of time. Some statues specifically proscribe such neglect of sick, or disable animals. One such statue penalizes the owner or an animal if any injured, sick, infirm, or disabled animal shall fail to receive proper food or shelter from said owner or person in charge of the same for more than five consecutive hours.

Interpreting this provision, courts must determine whether shelter will be limited to protection from the weather or will it also encompass protection from other animals due to such behavior as chicken cannibalism or hog tail biting. A very limited interpretation might limit shelter only as protection from the weathers. If the purpose of the regulation is to help an injured animal to recuperate, then shelter from other animals would be equally as important as shelter from the weather. No evidence yet exists indicating which interpretation courts will follow. Due to the lax enforcement of the neglect, it now appears unlikely a court will ever be forced to make such a determination.

The construction of state anti-cruelty statutes illustrates how ineffective they have been in stopping factory farm abuses. In the

Andrew D. Anderson

first instance, such statutes might specifically exclude farm animals from the definition of "animal." Second, the legal definition of cruelty requires a showing that the abusive practices are "necessary or unjustifiable" and in those statues, which delineate what acts are cruel or that the activity is one of a number of enumerate abuses. Even where a particular statue implied proscribes certain factory farming abuses, enforcement prodders against the violators create another serious obstacle.

No Enforcement of Laws

Private enforcement programs grant police power to local humane societies and allow court actions by private citizens. Most state enforcement programs primarily rely on criminal prosecutions, although private civil actions also are sometimes involved. While the impotence of these several enforcement programs ultimately results form the substantive deficiency of the laws to be enforced, the weak enforcement structure itself is instrumental.

Many states do not assign responsibility for enforcement of the anti-cruelty statues to any particular agency, relying instead on the local police or sheriff. Since this approach has failed to adequately prevent conventional animal cruelty offenses, it seems equally inadequate for preventing intensive farming abuses. Due to the increased incidences of other types of crimes, police officials simply are not equipped with the resources necessary to actively enforce animal protections statutes, but only respond to warrants sworn out by others.

Other states have created specialized bureaus of animal protection to perform such duties as secure the enforcement of the law for the prevention of the wrongs to animals and promoting the growth of education and sentiments favorable to the protection of animals. Such an independent bureau could be more effective in protecting farm animals, although the influence of politics and agribusiness might also restrict its effectiveness in arresting factory farm abuse.

Local SPCAs have generally avoided controversial areas like factory farming, abdicating the responsibilities of enforcement to concerned private citizens. Since cruelty to animals constitutes a criminal act, enforcement at the bequest of private citizens remains a virtually futile gesture. Because a private citizen cannot make arrests unless the offense is committed in his presence, such citizens' arrests for animal cruelty seldom, if ever, occur. Normally, private citizens contact the local police or SPCA and file a complaint, shifting enforcement responsibility back upon the very institutions whose inaction prompted the private citizen to act.

Under some statutes, if the complainant receives an inadequate response from these institutions, he may petition a magistrate to issue a search warrant authorizing the appropriate officer to investigate. This procedure vests a large amount of discretion in the magistrate, who may be hesitant to issue such an order. Moreover, the private citizen must undertake so much affirmative action that it is likely that few would actually preserve to the conclusion.

None of these enforcement plans can adequately implement cruelty statutes to protect animals from the brutality of factory farming. Traditional police protection is inadequate because of the priorities placed on other crimes. State departments of agriculture and state bureaus of animal protection remain ineffective due to the traditional agribusiness influence over those groups. Local choice to enforce anti-cruelty statutes, have historically avoided such controversial issues. The private citizen may act, but traditional notions of standing and the current non-recognition of legal rights for animals generally renders the private citizen powerless as well.

Even if there were effective enforcement agencies, our country lack workable anti-cruelty laws for them to enforce. At the federal level, some regulatory statutes protect non-farm animals, but none protect farm animals. State anti-cruelty statutes offer little more protection, since they too are not intended, constructed, or enforced to protect farm animals.

Fortunately, the bleak legal status of farm animals in the United States does not prevail throughout the world. The United States has just begun to recognize the injustice of factory farming. Other countries have actively attacked the problem through legislative reform and government commissions. Although the vast size of agribusiness in the United States makes reform in this country particularly difficult, much of the activity in foreign countries offers and excellent model for reform.

Models of Reform

In 1965, in response to the public outcry which followed the publication of a book depicting the deplorable condition of farm animals, England conducted perhaps the largest government investigation into the abuse of intensive farming. A committee consisting of experts in veterinary science, animal husbandry, and agriculture compiled a report at the conclusion of the investigation which not only reported on the treatment of animals in intensive farm operations, but also proposed reforms. The committee's proposals included such recommendations as,

1. Minimum space allowance for chickens and a prohibition of de-beaking.
2. Minimum space allowances for pigs and a prohibition of routine tail docking.
3. Prohibition of the confinement of sows.
4. Freedom of movement and a diet of iron and roughage for calves.
5. A general demand for better stockman ship.

Although these proposals did lead to Codes of Practice being issued by the Minister of Agriculture, many animal welfare activists have criticized the Codes as adapting too few of the suggestions of the committee report and for watering down those which were adopted. For example, the Codes sharply reduce the space allowance for poultry, permitted de-beaking, allowed slotted floors for cattle and continued the abusive practices of the veal trade. The most widespread criticism of the Codes is that they are merely government recommendations and do not have the force of law. Nevertheless, they do represent an important breakthrough in government regulation of intensive farming.

Other countries have been more effective in enacting reform legislation. In 1976, the French legislature enacted a Law on the Protection of Nature which permits the State Council to take measures in order to protect domestic animals from maltreatment and the sufferings resulting from the manipulations inherent in the various rearing methods and methods of transport and slaughter.

By referendum vote in December 1978, an overwhelming majority of the Swiss electorate accepted a new federal law that:

1. Requires the Federal Council to regulate minimum size and construction requirements of animal enclosures.
2. Set standards for keeping of piglets in battery cages.
3. Limit the keeping of farm animals in total darkness
4. Prohibits the keeping of calves on grid floors.
5. Prohibits the keeping of poultry in battery cages.
6. Requires that surgical procedures on animals be done by a veterinarian under general or local anesthesia.

Although this provides an excellent framework, the actual effectiveness of the law depends on the detailed regulations which were currently being prepared by the federal veterinary office.

In West Germany, the German Animal Protection Act of 1972 provides that, any person who is keeping an animal or who is looking after it:

1. Shall give the animal adequate food and care suitable for its species, and shall provide accommodations that take account of its natural behavior.
2. Shall not permanently so restrict the needs of an animal of that species for movement and exercise that

the animal is exposed to avoidable pain, suffering or injury.

The act's importance results from the first piece of legislation to explicitly recognize behavioral distress. West German law also authorizes the Minister of Food, Forestry and Agriculture to regulate tethering, cage size, feeding equipment, lighting, temperature, ventilation, care, and supervision by the farmer.

In Denmark, the caging of laying hens has been banned since 1950. Danish law forbids force-feeding, castration, and tethering so as to cause discomfort or pain. The laws of Norway and Luxembourg contain similar clauses. Also the 1965 Animal Welfare Act of Luxembourg prohibits the housing of domestic animals in such a manner that they suffer from the lack of space in the stall or enclosure in which they are kept or from inadequate ventilation, lighting or protection from the elements.

Sweden, the Swedish Animal Protection Act of 1944, provides that animals shall be treated well and as far as possible protected from suffering. The animal housing shall provide adequate space and shall be maintained at a satisfactory level of cleanliness. In addition, the law sets minimum space requirements for calves, hogs and chickens, and forbids the transport of calves less than two weeks of age.

European countries have led the factory farm reform movement, but such reform is more easily accomplished in smaller nations. Europeans traditionally have had a more humane regard for animals than Americans. Agribusiness in Europe has not attained as great an influence in decision making. In light of the different situation in America, reform in the United States must be very well planned for it to be successful.

Chapter 9: Are We Helping Ourselves?

Irrigation

Ten percent of the water consumed worldwide is for household use. Agriculture's takes 70 percent, and half or more of that water is lost to evaporation or runoff. Drip irrigation, which uses perforated tubing in the ground to deliver water to crops use 30 to 70 percent less than traditional methods and increases crop yields. The first drip systems were developed in the 1960s, but even now they are used on less than one percent of irrigated land. Most governments subsidize irrigation water so heavily that farmers have little incentive in drip systems or other water saving methods. Industry consumes the remaining 20 percent of water, often inefficiently. Reusing water and adopting other conservation measures could help the world industry cut its water demands by more than half.

"Twelve trillion" gallons of water were taken from the ground in 1950 and the figure more than doubled in 1980, and each day 21 billion more gallons of water flow out of water resources than flow in from rain or melted snow. We must remember, the past of the Middle East that is now a desert was heavily irrigated in Biblical times.

Irrigation brings the land into production, protects against drought, facilitates double cropping, and permits more profitable crop production. Its use has tripled since 1940 even though the total amount of land cropped has remained fairly constant.

Irrigation may affect water quality in one of a number of ways. Irrigation salinity levels may be higher than natural water flows because the irrigation water dissolves salts and exist in relatively dry soils. These salts in turn become further concentrated through evaporation-transpiration and may cause a build-up of salinity levels in the root-feeding strata where there is inadequate drainage. Irrigation water also may either deposit sediment in the receiving soils or cause additional erosion, and may carry nitrogen,

155

phosphorus, and pesticides to the receiving waters. Finally, irrigation may alter the hydrological characteristics of the receiving system.

Irrigation water may infiltrate into the ground water, or it may flow as tail water to surface waters as either a non-point or a point-source, such as a flow from a field drainage system. Irrigational impacts on water quality vary tremendously from one area to another, depending on such factors as initial water quality, soil composition, irrigation technique, agricultural practices, investments, weather, and climate.

The omen that is heavy irrigation may destroy the land by salt seepage and wipe out societies that grow up around the man made oasis. Three years ago, Senator William Armstrong Republican of Colorado warned, the 1,400 mile Colorado River is the lifeblood of 17 million people, from Denver to San Diego. This river has made America's western desert grow fine crops, in fact, 1.5 million acres of prime farmland are irrigated by it today.

Yet this magnificent river is being slowly poisoned as its water becomes more and more saline and adulterated by dissolved solids. Salinity is caused by two things, salt loading which comes from contact with the very saline western soils, salty mineral springs, and by salt concentration which is caused by evaporation and the increasing use of the river in the seven states it serves. The salt load of 10 million tons annually which enters Lake Mead adversely affects more than 10 million people and one million acres of irrigated land.

According to the United Nations Environmental Program, 90 percent of the land in Egypt, 68 percent in Pakistan, 50 percent in Iraq, 38 percent in Peru, 30 percent in the United States, and 20 percent each in India, Australia, and Russia are suffering Stalinization caused by irrigation. Sodium in the soil or irrigation water accumulates at the root level of soils and turns into a sterile, rock-hard crust. Many hectares of irrigated land are pulled from production each year because of water-logging and salinization, the result of poor land management.

Senator Dennis DeConcinci, Democrat of Arizona, explained to the National Press Club recently, what's happening to water in America is more than an occasional accident, or even a series of isolated problems. The problem is more insidious than that. We are not running out of water, or even destroying it in the military sense. Water is becoming unusable because a lot of it is being contaminated, both above and below the ground. Water is also becoming unusable because delivery systems are old and falling apart, especially in the West, and because they cannot be built fast enough to keep up with population shifts in the Sunbelt States.

The crisis is not just in the West. A study by the Army Corps of Engineers found that population growth is dangerously increasing salt levels in the giant Chesapeake Bay. The consumptive loss of fresh water, by drawing fresh water from the tributaries, will rise from 500 million gallons a day to more than 2.5 billion by 2020.

In culture after culture, from Sumeria to New Mexico, massive irrigation of arid lands follows a familiar pattern. First there is property, and the culture expands. But rather quickly the mineral salts in the irrigation water increase the salinity of the soil, and food production drops. Farmers try to wash the salts out with even more water, and while there is some success, ground water levels rise, surrounding vegetation changes, and soil erodes away. Cities and pueblos are abandoned, and civilizations which once flourished by irrigation vanish.

Some 4.2 trillion gallons of water reach the United States in the form of rain or snow every year. About 92 percent of this evaporates immediately or runs off unused into the oceans. We withdraw some 400 billion gallons per day to irrigate, power, and bath America, 65 percent comes from freshwater sources such as lakes, rivers, marshes, reservoirs, springs, 20 percent from underground aquifers; and 15 percent from saltwater sources, such as inland seas.

Irrigation and Efficiencies

Irrigation efficiencies can be increased more than 20 percent by using "drip or trickle irrigation systems that supply water and fertilizer directly onto or below the soil surface. Experiments with the drip irrigation in the Nevada Desert yielded an increase up to 80 percent over wasteful sprinkler systems. Our present sprinkler systems spread small amounts of water a few feet above the soil and it falls to the ground. Evaporation starts immediately, continue as it falls to the soil and rest on the soil until it soaks in.

A General Accounting Office study found that more than 50 percent of the irrigation water is wasted. Water is evaporated out of irrigation canals at a rate sometimes as high as 50 percent and many large scale irrigators use the huge center-pivot rigs that spray water into the air, instead of drip irrigation developed by Israel to save water sprayed onto the soil.

Normal irrigation is inefficient. As much as 70 percent of the water used may not even reach the crops. And a great deal of irrigation is directed at comparatively inefficient activities, such as growing livestock feed. Half of all the fresh water used in the United States is used for feeding and watering livestock.

In California, the livestock industry uses one-seventh of the state's water, but contributes only one five-thousandth of the state's income, according to water analyst Marc Reisner, who has studied water use policies in the Americana West. This is possible because the federal government has built dams and other projects that supply water to farmers and ranchers throughout the West at prices well below the water's real cost. In 1981 the United States Government Accounting Office reported that farmers who grew cattle feed with water from a $500 million project near Pueblo, Colorado, were paying seven cents for a quality of water that cost $54 to produce. If water was priced more realistically it would be used more sparingly, and irrigation could become up to 20 percent more efficient.

In summer the taps run only a day or two a week in Amman, Jordan, so residents have to store water in rooftop tanks. The rations will likely get even tighter in years to come. Jordan's population is on track to double within a quarter of a century. In Las Vegas, the driest state from rain fall with excess and illusion, water is no exception. At the Bellagio Hotel, 27 million gallons of water dance to show tunes through choreographed nozzles in an eight-acre artificial lake. With nearby Hoover Dam providing precious Colorado River water, Las Vegas residents all have pretty green lawns, but the city is beginning to realize in the year of 2003 that there is a growing problem of water shortages.

The first drip systems were developed in the 1960s but even now they're used on less than one percent of irrigated land.

A sixth generation Afrikaner, Schoeman presides over South Africa's largest family-owned citrus farm, an operation that covers 4,400 acres in the fertile, heavily irrigated Olifants River Valley. The Schoeman farm has 500,000 citrus trees that annually produce 175 million oranges and lemons for export to 32 countries. At the heart of this flourishing enterprise is a sophisticated irrigation system that farmers must change to as water becomes scarcer and more expensive.

Since joining the family business 27 years ago, Schoeman has helped introduce a succession of irrigation technologies. When he began, the farm simply opened the gates of irrigation canals and flooded the citrus groves, a highly inefficient system still common in the world today. In the 1980s more efficient sprinklers were introduced. Now Schoeman is steadily replacing the sprinklers with super-efficient drip irrigation, which gives the trees exactly what they need every day by parceling out small amounts of water to each tree. As Schoeman has used ever more efficient irrigation systems, the farm has quadrupled the production of fruit per acre while actually using a third of the water.

The nerve endings of the present system are yard-long computerized probes that the irrigation manager, Jaco Burger, places in the soil beneath tidy rows of trees. Every 15 minutes, via solar-powered radio, the probes relay data about soil moisture to the farm's computers. Based on that information and the time of year, the trees need different amounts of water during the different stages of fruit development. Burger adjusts the rate at which water, mixed with fertilizer, flows. Standing in a hundred-acre field, surrounded by about 35,000 young orange trees, you watched, as water trickled from a narrow tube into the soil below a sapling, one of three 20 minute feeding pulses the trees received that day.

Burger said commercial farmers will have to continue making such technological leaps as water becomes costlier. As he put it, farmers in the Olifants have been paying next to nothing for water. That and many other aspects of water in South Africa are beginning to change.

Farmers using the old irrigation system must look at the net financially gains of the two systems. Surely there will be a cost involved, but considering the net gain, the amount of water used, and the amount of water lost over a different number of years will provide a positive answer for the drip system.

Most governments subsidize irrigation water so heavily that farmers have little incentive to invest in drip systems or other water saving methods. Industry consumes the remaining 20 percent of that water inefficiently.

Just because the water's there does not mean people can use it. Underground aquifers have a hundred times more water than lakes and rivers, but most of that water is too deep to reach. Shallower aquifers are being quickly overdrawn in many parts of the world, and much of Earth's surface water either rushes to the sea in floods or ends up in places far from the people who need it. For instance, Canada has a tenth of the world's surface fresh water but less than one percent of its population.

Migratory birds still flock to California's Tule Lake, but the lake is little more than a 10,000 acre puddle compared with a century ago, when it reached ten times that size. Most of the water has been delivered for agriculture. In a trend echoed worldwide, 90 percent of California's wetlands have disappeared, and 39 of 67 native fish species are extinct or at risk of extinction. On the Klamath near by River in Oregon, federal efforts to save three fish species led to a confrontation with farmers last year when officials shut off irrigation water during a drought. It was later restored.

Agricultural policies now in place define the very idea of unsustainable development. Just 15 crops such as corn, wheat and rice provide 90 percent of the world's food, but planting and replanting the same crops strips fields of nutrients and makes them more vulnerable to pests. Slash and burn planting techniques and over reliance on pesticides further degrade the soil.

Solving the problem is difficult because of the ferocious debate over how to do it. Biotech partisans say the answer lies in genetically modified crops, foods engineered for vitamins, and yields and robust growth. Environmentalists worry that playing about with genes is a recipe for disaster, but there is no reason that both methods cannot make a contribution.

Better crop rotation can help protect fields from exhaustion, salt, and erosion. Old fashioned cross breeding can yield plant strains that are hardier and more pest-resistance. But in a world that needs action fast, genetic engineering must still have a role, provided it produces suitable crops. Increasingly those crops are being created not just by giant biotech firms, but also by home-grown groups that know best what local consumers need.

Making Good Thing Better

Macleod decided to take about 10 million of the 125 million gallons of wastewater the city treated daily and use it again, piping it to industries nearby. The French firm Vivendi built a sophisticated treatment facility next to one of Durban's wastewater plants. Operators in the nearby paper mill and refinery were satisfied because they paid almost half price for the recycled water and it cut metropolitan water demand by about 5 percent.

The waste passed through sand and carbon filters, treated with ozone, mixed with chlorine in twelve hours, and its taste was no different than municipal drinking water and the treated wastewater. It is intended for industrial use.

Vivendi is in the neighboring desert nation of Namibia and a partner in a plant that turns wastewater directly into drinking water for the capital, Windhoek, refining the water even one step further than Durban's recycling operations. Windhoek's wastewater-to-drinking water plant is the only such facility in the world, but Stephen McCrley, the general manager of Vivendi's Durban operation, is confident it will not be the last.

Dams Do Cost

The reach and benefits of dams are enormous. In 1950 there were 5,000 large dams worldwide. By the year 2000 the number had grown to 45,000 large dams that caught 14 percent of all precipitation runoff, provided water for up to 40 percent of irrigated land, and gave some 65 countries more than half their electricity. But the cost was also enormous. In India alone, up to 38 million people have been displaced by large-scale dams.

Large dams alter the flow of rivers and drown land with reservoirs, interfering with fish migration and flooding cultural sites. Three Gorges Dam on the Yangtze River may displace nearly two million people and flood an estimated 240,000 acres of cropland when completely operational in 2009. The Rio Grande dam at several points in the United States and Mexico last year, ran dry at its mouth. Turkey dams on the Tigris and Euphrates River have already begun to affect Syria and Iraq threading disaster. How well we manage our water all over the world will write the story of the 21st century.

In Gujarat's largest city, Ahmadabad, the Sabarmati River once flowed perennially through the heart of town. Today, due to the construction of a large dam and over pumping of the region's aquifers, the river only runs during the monsoon floods in summer. The rest of the time the Sabarmati is a dry dusts shrouded scar inhabited by tens of thousands of people living as squatters in shacks.

United States Sharing Water in North

The Army Corps of Engineers wanted to divert water from two large reservoirs in South Dakota to bolster the Missouri River's sagging water levels farther down stream of 2,341 mile Missouri River. A federal judge stopped them. They went to North Dakota and Montana to divert some water from the lakes and reservoirs there, but another judge stopped them. Another judge in Lincoln, Nebraska ordered the corps to either raise the river's water level or face a contempt citation.

Later the Army Corps finally succeeded in draining some water from Lake Sharpe and Lewis and Clark Lake, which are two medium-size Missouri River reservoirs in South Dakota. The corps were continuing their search for more water to ship south amid a bitter regional battle.

The upstream forces of South Dakota were led by Senate majority leader Thomas A. Daschle fighting to protect the $85-million fishing and recreational industry around the Missouri River reservoirs and lakes. Downstream interest was led by Senator Christopher Bond of Missouri trying to maintain sufficiently high river levels for the aging and far less lucrative barge industry and to provide municipalities and nuclear power plants with a dependable supply of water.

The real problem was the corps wanted to divert water from the upper Northwest to the Midwest and when North Dakota and South Dakota were experiencing their third year of drought, while Missouri and other Midwestern states down stream were being pounded by heavy rain and flooding. Only a fraction of the lower Missouri, much of it engineered into a straight concrete-reinforced channel, was experiencing flood conditions while other upper sections were below normal levels.

Five hundred miles of the river was not flooding and was just above what was necessary to float the tow boats that went down that part of the river. Yet the government was trying to shift water from a drought-stricken region to a flood zone suggested to some that they did not know just what they were trying to do.

Both senators urged President Bush to intervene. North and South Dakota officials said that by diverting water, the corps would lower lake levels even more in the midst of the spawning season for walleye, a popular game fish, and smelt, a bait fish. South Dakota obtained a court order temporarily protecting water levels of Lake Oahe and Lake Francis Case and a few days later a North Dakota judge did the same for Lake Sacagawea. The third judge in Nebraska ordered the corps to maintain sufficient water flow in the lower Missouri to accommodate barge traffic.

The Army Corps of Engineers unveiled options for alternating the management of the river, including one to use dams to manipulate its flow to mimic spring rise and summer ebbs as a way of giving endangered fish and birds a chance to survive.

California Demands Water in West

Imperial's Valley poorest county uses 1 trillion gallons of river water every year to transform a desert into one of the richest farm belts on earth and refuse to sell a drop of their water to be used in California. The Imperial Valley sees the water as their birthright. Valley farmers in the 1900's were the first to tap the Colorado River using mule teams to dig canals.

Imperial County is a flat expanse of desert, sandwiched between two mountain ranges East of San Diego. El Centro is the county seat with low-slung stucco buildings sitting amid lush rows of alfalfa, lettuce, and other crops. Half of all jobs are tied to agriculture, many of them held by Hispanic farm workers. Today the valley is the nation's largest irrigation project, producing about $1 billion worth of cattle and crops each year.

The number one water consumption is irrigated pastures, grass and hay for cows and sheep, and in 1986 used 5.3 million acre-feet of water. That was equal to the amount of water consumed by all 27 million people in the state including swimming pools and watered lawns. In that same year, irrigating alfalfa required as much water as metropolitan Los Angeles and the Bay area combined including drinking water, lawns, gardens, toilets, showers, swimming pools, and car washes. Alfalfa returned about $630 million which is less economic activity than a few square blocks of downtown Los Angeles.

The third water thirsty crop grown is cotton. The San Joaquin Valley's million acres of cotton required in 1986 around 3 million acre feet of water per-year. That was enough water for 15 million people or 15 times the population of Nevada.

Another crop that is water-consumptive is rice grown in the valley. That is really grown because many of our rice farmers buy water from the bureau of Reclamation at fabulously subsidized prices, as well as for cotton, hay, and alfalfa. Rice requires large amounts of water since it grows in the desert in manmade lakes. The valley used much more water than the whole Bay Area and enough for 10 million people and the gross value was a small $204 million.

It has little in the way of money, jobs, or people. The county's 145,000 residents have the state's lowest median income and the highest unemployment rate. The only things they do have are massive amounts of land and water. About 70 percent of the state draws water from the Colorado River that flows through the valley.

Residents were upset at the deal that called for waving some farmland idle to free up water for upscale San Diego. They have the fear that it will suffer the same fate as Owens Valley, a California farming community that saw what remained of its water and its future flow down an aqueduct to Los Angeles 90 years ago.

The water-sharing deal hinged on Imperial County selling a portion of its Colorado River allotment to Los Angeles and fast growing San Diego. The Colorado River is harnessed by the massive Hoover and Glen Canyon dams and serves 25 million people from Denver to San Diego. Under pressure from residents that want to preserve jobs and tradition, county water officials said no.

The Interior Department that oversees the lower Colorado said as of January 1, 2002 it will withhold from California enough river water to supply 1.6 million households. To that, Southern California water officials said they have enough reserves to offset the cut for the region's metropolitan areas for at least two years.

Politics entered when Bennett Raley, the Bush administration's point man on Western water issues said Interior Secretary Gale Norton could use her authority to take water away from farmers who are found to be wasting it.

Neil Grigg, a water expert at Colorado State University, sees Imperial Valley as the most pivotal irrigation district in the whole country. The transfer of water to cities is bound to happen, he said. The district is going to have to look at what is in everyone's best interest.

For most of us, the answers to our water problems are to be found in the reuse of urban and industrial water, of conservation and careful use in agriculture, desalination plants, and restoration of watersheds.

The desert resorts in Palm Desert, California located in the Southern California desert, is also having water problems. How much longer can the people at the resort water-ski on a man-made lake, travel by gondola to waterfront bistros, and golfers tee off at more than 100 courses made lush, pretty and green from constant watering.

The federal government in 2003 cut the amount of water California can draw from the Colorado River that will hurt the future of the resort and retirement Mecca 110 miles east of Los Angeles known as Coachella Valley. The drought and booming growth around the West prompted the government to crack down and demand that the state's water agencies work out a deal to redistribute the water. This is necessary because California has long been using more than its share of water from the Colorado River with also flows to seven Western states.

The government cut back the state's share of river water by 15 percent after a deal fell through December 31 and the major portion of that cut was in the Coachella Valley. This lead to a cut back by the valley's water agency that halted deliveries of Colorado River water to about a dozen golf courses, the lake built for water skiing amid a housing development, and one construction company. Also landscaping ordinances have required new developments to use 25 percent less water than existing ones.

The Trilogy Golf Club at La Quinta is only one of several spending more than $200,000 each to drill into the aquifer far beneath the course. The proper people have been warned because the farms, homes, golf courses and other resorts that use well water are sucking so much from the ground and for so long that the valley floor sinks more than an inch a year in spots. This will accelerate if they keep using well water and if more water cannot be gotten from the Colorado River.

John Wesley Powell was a 19th century explorer and said, the desert was the most desolate region on the continent, but the cheap and abundant water from the aquifer transformed it into a landscape of luxury neighborhoods decorated with waterfalls and lakes and fairways. But the people did not know or seem to care that the water might run out in future years with heavy usage. It is a 300 mile valley that stretches from what was former called the Rat Pack getaway of Palm Springs. It became popular in the 1950s and the population grew 170 percent between 1980 and 2001to approximately 330,000.

In 2002, the golf courses helped attract 3 and one half million visitors that spent an estimated $1 billion.

There is another problem that consist of fruit and vegetable growers that use most of the valley's Colorado River water allotment. They are paying $15 million over a five year period which is nearly 10 times the usual cost to buy excess water from farmers in the nearby Palo Verde. However, water officials are hoping that the valley and three other Southern California water agencies reach an agreement to share the Colorado River and get enough water to supply farmers and recharge the aquifer for at least the next 35 years. There does not appear to be any interest in using less water or that the water will be sucked up completely.

The Vanishing Water

Israel and Jordan announced their largest joint project ever that will consist of an $800 million pipeline intended to save the shrinking Dead Sea from environmental devastation. Shared between two countries, the sea level is sinking at nearly one yard a year and will disappear in a few decades that would drain scarce water supplies in the region, and hurt tourism.

The two governments hope to build a 190-mile-long pipeline from the Red Sea through both countries to halt the decrease in water level in the Dead Sea. Jordan and Israel signed a peace accord eight years ago but the agreement did not work out. Bassem Awadallah, Jordan's minister of planning said, nature knows no boundaries and no political conflicts.

The Dead Sea is the lowest point on earth, at 400 yards below sea level and is the saltiest large body of water in the world. It also has unique minerals used for health treatments. Its potash fuels a major chemical industry, and its beauty attracts thousands of tourists. Much of the water from the Jordan River, which ends in the Dead Sea, has been diverted for use in the region. It has begun sucking up vital sources of fresh groundwater, causing massive sinkholes to appear on both sides of the border, and Jordanian officials said they had to evacuate 3,000 people because of the sinkholes. Officials have nicknamed the pipe, "peace conduit," that will end the Dead Sea's decline and slowly restore it to an ecologically appropriate level.

Both countries are planning for the future. They want to build a canal and a desalination plant that will provide fresh water for Jordanians, Israelis, and Palestinians. It is estimated to cost $3 billion and would take more than a decade to finish. Awadallah said, "the quicker we all turn our attention, our resources, our time and effort toward construction, and not destruction, there will be guaranteed justice."

The South African government has succeeded in bringing safe drinking water to millions of its citizens in recent years, but about 7 million of the country's 44 million people still lack access to clean water within 200 yards of their homes. Drinking contaminated water, often from streams, was behind the most recent outbreak which killed 289 people in Kwa Zulu-Natal and infected 120,000 others since August 2000. Those deaths are among the 18,000 that occur in South Africa annually from diarrhea-related ailments, most of which were waterborne.

The outbreak was centered in the hilly, picturesque Ladysmith region, where whites typically have water and sewage systems, but where 85 percent of blacks lack proper sanitation, and 60 percent do not have access to the South African government's minimum recommended quantity of clean water, 25 liters, 6.6 gallons per person per day.

One city in the United States is preparing for the future, St. Petersburg, Florida. It is stocking parks with drought-tolerant plants and recycling its waste water. Used for irrigation and industry, treated waste water may one day meet a fifth of the city's needs. Worldwide, two-thirds of urban waste water does not even get treated, much less recycled, but that could change. Santiago, Chile, which just opened a treatment plant, will treat all its wastewater by 2009.

It could happen anywhere. In the Castil la Mancha region of south central Spain, a 74 year old former fisherman has seen one of the country's prized wetlands altered beyond recognition. He fished in the freshwater marsh at the heart of a sprawling 60,000-acre mosaic of wetlands in La Mancha. He fished on the picturesque Guadiana River and remembers an area called "Los Ojos-The Eyes," where large underground springs bubbled up into the limpid waterway. I would sit in my boat and see the water rising up in big columns. Now it looks like the moon.

Los Ojos is no more. The underwater springs dried up in 1984 and a six-mile portion above Daimiel also has disappeared where 30 yards of percolating river was. Now there is a road, fields of grain, and rocky portions of a riverbed. The 60,000 acres of original wetlands, superb habitat for cranes and waterfowl, have shrunk to a core area of about 24,000 acres.

La Mancha has witnessed an explosion of well digging in the past 40 years that has lowered the water table and reduced stream flows. On the number of irrigated acres, farmers grow alfalfa, barley, corn, wheat, and sugar beets that has soared from 60,000 in 1960 to 500,000 today. The number of wells has grown from 1,500 to an official count of 21,000. Some experts say the total number, including illegal wells, could surpass 50,000.

As long as you have so many wells sucking out the groundwater, Las Tablas won't come back. As the world's population increases and food demands soars, getting more out of each drop of water is imperative. Another "two billion" people will need food and water by 2025. Will the planet have enough water? That depends on how wisely it is used.

Spain is at a crossroads like Southern California, but California does not know it. The Mediterranean country has built dams about 1,200 major ones, and piped water long distances to supply farms and municipalities. Now a new national Hydrological poplin calls for transferring nearly 1.4 billion cubic yards of water a year from the Ebro River in the north to burgeoning regions along the Mediterranean coast.

Large numbers of people in Natwarghad village in the Indian state of Gujarat gathered around a large diameter hole dug into the ground. They were waiting for hours with ropes and containers on the ends going into the hole to pull water up when it slowly began to collect. All around village ponds thirsty people waited for water to be brought in by tank trucks paid by the state as the temperature was 111 degrees F with no shade in sight.

"Millions of people" in northern China face water shortages this summer as the Yellow River falls to its lowest level in 50 years, and environmental official warned. More than half the watersheds of China's seven main rivers are contaminated by industrial, farm and household waste an official said in a bleak annual report on the nation's environment. Xie Zhenhua is head of the State Environmental Protection Administration and said, "China is a country that lacks water resources, and problem of water pollution remains severe."

Blooming and growing industries and a population of 1.3 billion people have outstripped china's rudimentary water and sewage systems and left its cities choked with smog. Two-thirds of China's cities still is considered polluted by official standards. Only one-quarter of the 21 billion tons of China's annual output of household sewage is treated. Treatment plants are being built, but will still handle only half of all city sewage, leaving rural waste water untreated and the population keeps growing. The government has forecasted an annual water shortfall of 53 trillion gallons by 2030 which is more than is consumes now in a year. That figure looks like this, 53,000,000,000,000.

More problems are being created by drought and overuse that left the Yellow River so drained that in recent summer low sessions it has dried up before reaching the sea. Oil spills and water shortages on the river forced the government to suspend work on a project to divert some of its water further northward, said Wang Jirong, a deputy director of the environmental agency.

The 3,415-mile Yellow River winds its way from the mountains of western China to the Bohal Sea in the east, providing water to 12 percent of China's population. However, the monitoring of 185 sections of the river showed the water quality was poor with nearly half the stations recording pollution levels below grade V. That grade is the poorest measurement China has the Xinhua News Agency reported. Pollution, erosion, overgrazing and other forms of environmental degradation are taking a heavy toll on the health of China's people and on the nation's s economy.

Donna of the Atlantic coast is home to half a million wintering birds and a stopover for six million migratory birds, has seen its natural marsh lands cut from 370,000 acres to 75,000 because of agricultural development and water engineering projects. It still attracts large numbers of flamingos, white storks, glossy ibises, grey lag geese, and other water fowl because the wetlands were converted to flooded rice fields or ponds. It is now working on an eight-billion-dollar plan similar to Florida Everglades to restore water flows that will revive some of Donna's marshes.

How do we balance human needs with the requirements of natural systems that are vital to sustain life on earth? Some are hoping that new technologies, such as the desalination of sea water, will solve the problems faced by a water-stressed world. Yet only two-tenths of one percent of the water people use today is desalinated, and most of that is produced in desert kingdoms and island nations. Plants are under construction in California and Florida but they are costly.

Chapter 10: Paradox of Technology

The Green Technology

Technological developments come about as people seek solutions to specific problems and needs, and they open the way to other negative innovations and applications that were unimaginable at the beginning. Nearly all developments carry with them the potential for misuse and the consequences are unfortunate because the intended solutions will meet our requirements. New discoveries or technology will be applied in ways that transcend the intentions and the purposes of its inventors and new technology will reveal consequences that were not anticipated.

The development in agriculture improved food production around the world. Chemical fertilizers contributed to the increase in the world's food production greatly increasing the economies. Countries like India for decades were unable to feed its people, but became net exporters of food. At the same time, growing reliance on insecticides and fertilizer contributed widespread chemical pollution of rivers, lakes, and seas threatening the food chain.

The world had never seen such an improvement, but it did not anticipate the environmental costs. The green revolution led to waterways deadened by eutrophication, ecosystems altered because irrigation had drained rivers and streams, and various real problems due to increased use of pesticides.

The development and use of pesticide on an insect that is resistant to a pesticide is virtually automatic, because those insects that survive are unaffected by that chemical and will breed off spring that are also immune, creating super bugs. The first use of insecticides reduced crop losses, but over time, the pests rebounded. The world wide number of pest resistance to chemicals continues to climb. Today, many hundreds of pests, including most of the world's major pests, are immune to existing pesticide.

Major new technological developments produced changes that deeply effected society and did so in ways that made it impossible to contemplate turning the clock back by rejecting the development. The very power and perceived permanence of new technology surely contributed to the wariness with which is regarded by man. The Green Revolution is a good example of the characteristic. New technologies can be adapted to address some of the unfortunate consequences of modern agricultural methods and abandon those methods. It would lead to malnutrition and starvation on a scale unknown in human history.

New technologies have encouraged the feeding of a wider range of materials to cattle and that also includes "waste" including poultry manure and it is shipped all over the world. This could be an opportunity for food borne pathogens to enter the food supply and spread. Farmers have shifted the diets of beef cattle from hay to grain in order to boost growth rates and reduce costs. James B. Russell and Jennifer Rychlik of Cornell University say, the microbial ecology is altered and the animal becomes more susceptible to metabolic disorders and infectious diseases.

Since drugs are effective in accelerating the growth of animals, farmers have been adding antibiotics to animal feed. Alicia Anderson, an epidemiologist for the National Antimicrobial Resistance Monitoring System said, the use of antibiotics as food supplements for farm animals is a serious threat to human health. She also believes the drugs in healthy animals are playing a role in changing the very nature of food borne bacteria creating strains that is resistant to antibiotics used in human medicine.

Chapter 11: Temperatures Rising Worldwide

Changes in The Great Lakes

The world's largest freshwater system has shrunk before, but never so quickly. In Traverse City, Michigan, empty chaises, chairs, at a resort on what once was lake bottom, reflect how the Great Lakes tourist economy has slipped in sync with falling water levels. The big new hotel sits abandoned and the farther the waters recede, the higher anxiety rises.

The waterline along Georgian Bay, Lake Huron, Canada, has dropped more than 40 inches in recent years, a legacy of drought and warmer temperatures. Sports fishermen fear for the future of species that spawn in tributary streams.

The level of The Great Lakes has fallen once again, all the way from Duluth, Minnesota, to Kingston, and Ontario, at the head of the St. Lawrence river. This is not just a matter of inconvenience to a hundred thousand landowners along United States and Canadian shores, though more than a few of them are being put to the expense of extending their docks. It is a matter of concern to the multitudinous cities and farms dependent on lake water, to the boating and fishing segments of the region's multibillion-dollar tourism industry, and to the operators of deep-draft ships that ply these inland ports and waterways to hitch North American's heartland to the markets of the world.

The Great Lakes, Superior, Michigan, Huron, Erie, and Ontario along with the rivers, channels, and lesser lakes feeding or draining them, constitute the largest surface freshwater system on earth. The system is spread across more than 94,000 square miles and drains a much larger watershed that embraces parts of eight states and two Canadian provinces. If the earth were flat and the lakes adaptable as buckets, there would be enough water to flood all the land of the Western Hemisphere under two feet of water.

Andrew D. Anderson

A hydrologic cycle orchestrates the volume of water contained in this system at any one time and the volume dictates the levels. The cycle begins in the clouds. Rain and snow fall across the lakes and the surrounding watershed lands, where runoffs should replenishes the system's tributaries and aquifers and they in turn replenish the lakes. But that is only half of the cycle. The other half takes some of the water away through evaporation from the surface of the lakes and transpiration from terrestrial plants through the watershed. When the inflow from precipitation and runoff is exceeded by the lost of water due to evaporation, transpiration, and outflow down the St. Lawrence River, the levels of the Great Lakes have nowhere to go but down.

Evaporation seems to be winning. By most accounts six of the warmest years on record in this region occurred in the past decade. That not only increased the rate of evaporation in the summertime but also raised it in the winter by depriving the lakes of their normal ice cover. Ice inhibits evaporation. With the exception of Erie, the shallowest of the five, the Great Lakes rarely freeze shore to shore, but often ice up in their bays and mid-lake areas. In recent years, ice cover did not occur in some places accustomed to freeze or, if it did occur, came in later and went out earlier than usual, which raises the question of global warming.

Lake levels are also hugely influenced by the annual snowcap and subsequent snowmelt runoff, especially in the headwater country of Lake Superior. Over three of the past five years, snow packs around the three upper lakes have yielded runoffs significantly below average. Springtime has been starting six weeks earlier than normal around the northern lakes, said Roger Gauthier, a senior hydrologist with the United States Army Corps of Engineers in Detroit. The only snow we have seen this year has been lake effect snow-moisture that evaporates off the lakes and then falls as snow, sometimes outside the basin.

An icebreaker met no resistance as it crossed Lake Erie toward the lights of Cleveland, Ohio. Normally in winter, sheet ice would be edging out from the shore. But ice has been rare in recent years, a chill omen for the future of the lakes.

The port cities of Duluth, Minnesota, and Superior, Wisconsin with a combined population of 118,500, moves up to 40 million metric tons of cargo eastward through the Great Lakes every year. The levels of the lakes are very important for moving cargo.

Last year the aggregate level of the five Great Lakes withheld a fifth of the world's surface fresh water, plunged to its lowest point in more than 30 years. Superior and Huron hit near-record lows. Despite slight increases recently, the aggregate level is still down. Scientists attribute the loss to a new cycle of dry, warm weather across the region's 200,000 square mile water shed.

Temperatures Are Changing

It is difficult to realize, but only a little imagination is needed when you see the beaches of some cities. All of the big hotels in Miami are built a few feet from the edge of the ocean with an assurance there will never be any danger. The beaches at the Cape on Martha's Vineyard are being continually washed away, and big hotels in Atlanta City, New Jersey, has stores and gambling building built on the edges of the beaches. There was a storm and high winds that caused major damages as water stood in the basements of the buildings.

The Atlantic Ocean widens up to 1.5 inches a year and the Red Sea widens 0.8 of an inch in a year. That is not much, but consider if increasing numbers of icebergs keep breaking loose and more snow on mountains keep melting, there will be a difference as they melt and increase the amount of water. This is another reason we should stop increasing the growing of pollution by riding trains.

A report from researchers in 99 nations that met in China said, increased floods, killer storms, and droughts are caused by global warming. The global temperatures could rise 10.5 degrees Fahrenheit in the next century causing more bad weather and flooding. Since the 1980s, temperatures have kept rising. In November of 2000, there were talks on how best to cut emissions from power plants, factories, and vehicles that causes heat in the atmosphere. If emissions are lowered, oil use has to be decreased.

A few years back, the undeveloped countries were asked to stop their pollution as well as the United States. Some of the comments were, the United States has already developed their country, yet they want countries like China to reduce their emission when they are building up their country and would emit more. Emissions per person increased 3.4 percent between 1990 and 1997 and the number of persons also increased causing a double edge.

Barrette Rock, coordinator for the federally mandated New England Regional Assessment said, we could have less colorful foliage and continued coastal storms. The manager for Maple Grove Farms of Vermont that makes candy and syrup said, maple syrup from the trees has started sugaring earlier. Texas had record setting days of 100-degree F, and glaciers in the north have started melting. Most climate watchers agree that the temperature increase worldwide in the last century is partly caused by human actions burning oil and other fossil fuels. When we burn gasoline in cars, wood for heat, and forest burn because of lightning, the gasses are trapped in the Earth's atmosphere and re-radiated toward earth making a warm blanket. One hundred scientists meeting in Shanghai said, new evidence shows the 21st century temperature increases could be even more dramatic than the 1.8 to 6.3 degrees Fahrenheit predicted by the Inter-Governmental Panel on Climate Change.

The Panel said that a more modest temperature jump could increase temperatures and the worldwide sea levels could climb to 17 inches. If that happens, all low-lying areas could be under water including parts of Nantucket. As we hear the figures and what is perceived as causes, we must realize that it will happen, it is just a matter of when.

People in Iceland said the winter's are not as cold and the permafrost is not as deep as it has been. The weather starts getting warmer sooner than what they call normal based on past years with the continued result of global warming, the oceans will continue to get warmer, more and more giant icebergs will be seen, and this could be problems for the shipping lanes.

More Heat from Burning Forest

Natural sources of smoke originate from the burning of grasslands and forests each year owing to lightning strikes and other causes. It has been estimated that in African Savanna, between the equator and the Tropic of Capricorn, more than 600 million tons of grass and low forests are burned each year during the dry period, and in the United States, the United States Forest Service estimates that an average of 120,000 forest fires occur annually, burning 600,000 acres of forest.

Since 1970 there has been widespread evidence of a decline in the growth of evergreen forests in the eastern part of the United States and Western Europe. In Europe only, 8.5-10 million acres of forests show signs of deterioration. In 1982 Germany federal minister of food, agriculture, and forestry estimated forest damage at 1,300,000 acres, about 8 percent of the country's forests. In 1983 a new study found that 6,250,000 forest acres were affected, 34 percent of the total.

In West Germany three quarters of the fur trees are affected, and damage to spruce and pine rose from 9 to 41 percent in less than five years. These three species compose two thirds of West Germany's forests. In Czechoslovakia some 500,000 acres are damaged and in the Erz Mountains 100,000 acres are reported killed. In Krokonose National Park near Prague 85,000 acres of spruce are dying, and the species has stopped regenerating in the mineral soils. In Poland 1,200,000 acres are affected.

In Katowice, forests researchers have reported that fur trees are dead or dying on 450,000 acres and the spruce forests around Rybnik and Czestochowa, also the other industrial region, are completely gone. Environmentalists had warned Poland that as many as 7,500,000 forests acres would be destroyed by 1990 if the present industrialization plans, based on the burning of high-sulfur "brown coal" was continued.

The local and regional smoke pollution effects of such fires can be serious. A forest fire may last for any number of days and shroud the area downwind in smoke so dense that breathing is difficult. In the summer of 1987 a large number of isolated forest fires, most of which were set by lightning strikes, burned in the Northwestern United States and covered the area with dense clouds. In the fall of the same year, from October 27 to about November 10, there were 9,000 forest fires reported in 13 states, from Virginia to Texas, that make up the Forest Service's Southern Region, and the total acreage burned was equal to 500 square miles.

In West Virginia, the smoke was so thick in some southern counties in early November that visibility fell below 100 feet. Smoke from fires in Kentucky and West Virginia was carried by wind currents into New England, Bridgeport, and New Haven, Connecticut, about 500 miles from Southern West Virginia. Visibility was only about 1.5 miles. Radio and television stations advised people to stay indoors and use air conditioners to reduce the irritation caused by smoke. People with heart and lung conditions were advised to take extra caution.

The smoke from forest fires is limited to the lower layers of the troposphere so the first rains will cleanse the atmosphere and the wind will dissipate the smoke. Therefore, a region can have a high density of smoke pollution and within a short time be free of it. If the wind is not blowing or there is no rain, the smoke will remain overhead for a length of time.

A forest fire in Canada caused heavy smoke over New England the last of June for two and one half days. Not even a trace of the sun could be seen. Prior to that fire near Cibecue Arizona, a 468,000 acre blaze destroyed ponderosa pines causing 30,000 people to be evacuated, and over 423 homes were destroyed. Federal prosecutors accused Leonard Gregg, 29, of starting the blaze because he wanted to earn money as part of a fire crew.

Workers were battling to save sequoias. The wildfire blackened more than 50,000 acres, half of the Giant Sequoia National Monument, in a region that had little or no rain since spring. Flames were within 2 miles of a grove of sequoias called the Trail of 100 Giants, a grove of majestic sequoias that are among the largest and most ancient trees on earth up to 220 feet tall and 1,500 years old. They include 125 giant sequoias over 10 feet in diameter. They can live more than 3,200 years and their bark is capable of enduring countless fires. But fires kill them when other trees such as cedar, pine, and fur spread flames to the sequoias' limbs, high above the ground. At 275 feet tall it is considered the nation's largest tree and the world's longest living thing based on volume.

It was burning in every direction, forester Lewis Jump said, and was the worst in the afternoon. The hot canyon winds were coming up and creating quite a bit of turbulence. At least 10 structures burned, and about 200 homes were threatened. The United States Forest Service said, the blaze was sparked by an abandoned campfire.

A 1,800 acre fire in California's Ken County was fully contained in 2002 after burning about 40 structures near Lake Isabelle. Young people playing with matches are believed to have started it. Lighting during a night in Oregon started new fires in the Cascade Range and the high desert, forcing the evacuation of a church camp outside Sisters. About 250 firefighters who had been assigned to the 4,400 acre wildfire were sent elsewhere. In Washington, a 25,000 acre fire on the north shore of Lake Clelan was 30 percent contained, but hot dry weather was forecast and evacuation notices remained in effect for 200 homes.

In Pine Flat California a 38,000 acre blaze roared through the deep valley of the Giant Sequoia National Monument. The tremendous amounts of heat generated by the fires were spread in the atmosphere increasing the temperatures.

Carbon Dioxide Is Trouble

Carbon dioxide also increases temperatures. A number of lakes in the world contain huge volumes of carbon dioxide and other gases such as methane. The world's "gassiest large lake," Kake Kivu, is in the Great Rift Valley of East Africa between Rwanda and Aaire. The water of the lake, 60 miles long and 1,300 feet deep, is charged with more than 63 billion cubic meters of methane and five times as much carbon dioxide. The methane gas in the lake is generated from bacteria in the sediments and the carbon dioxide seeps slowly into the lake from nearby volcanic hot springs.

The devastating effect of a discharge of carbon dioxide from a lake was dramatically shown in August 21, 1986 when Lake Nyos in the West African country of Cameroon discharged some 1.3 billion cubic yards of carbon dioxide into the atmosphere. A water surge accompanying the gas as it gushed more than 250 feet into the air and the lake was stained a deep red from an iron compound that was carried from the lake bottom by the escaping gas. The silent cloud of dioxide moved ten miles down the valley, 1,700 people and hundreds of animals were killed, and 4,000 to 5,000 people left the area.

When working normally, the circulation cycle brings warm water up for the tropics where it merges with the eastward moving Gulf Stream and travels up the coast of Europe. Easterly winds warm up as they blow across the current, keeping some European cities relatively warm. Without this effect, temperature in Rome would have a climate very similar to chilly Boston because the two cities lie on the same latitude.

Man made emissions:

1. Arsenic from coal and oil furnaces.
2. Benzene from refineries and motor vehicles.
3. Cadmium from coal and oil furnaces, burning of waste, and from smelters.

4. Carbon dioxide, from burning fossil fuels.
5. Carbon monoxide from burning coal and oil, smelters, and steel plants.
6. Formaldehyde from exhaust pipes of motor vehicles and chemical plants.
7. Hydrogen chloride from incinerators.
8. Hydrogen fluoride from smelters and fertilizer.
9. Hydrogen sulfide, from industrial plants such as refiners, pulp mills, and sewage.
10. Lead from motor vehicles and smelters.
11. Mercury from coal, oil furnaces, and smelters.
12. Sukucib tetra fluoride from chemical plants.
13. Sulfur dioxide, from coal, oil furnaces, and smelters.

Excessive air pollution can be created at all geographical levels from local to worldwide. The length of the incident will depend on a number of actors. A strong surface wind movement will disperse the pollutants relatively quick. If the pollutants reach the stratosphere, they may be carried around the world before they disappear from the atmosphere.

Sea and Ocean Levels Rising

Approximately 10,000 icebergs float in the ocean each year with the tallest seen equal to a 50 story building with only 10 percent visible above water. They usually travel up to 27 miles per day and melts at a rate of 10 feet per day.

Scientists have discovered and iceberg they named Godzilla. It was the largest iceberg ever seen by human eyes. This crashing behemoth, whose slow speed and size could have endangered shipping lanes that carried supplies to McMurdo research base in Antarctic. It was a dangerous dynamic iceberg that had broken off a larger block of ice and was shedding huge pieces of ice since its birth during the month of March in the year of 2000. In the year of 2001, it was less than half its original size and about twice the surface area of Rhode Island, and nearly as long as the state of Massachusetts. As it continued through the turbulent sea, it sheds huge slabs of ice groaning, cracking, and heaving.

In the ocean, the action is at the edges for biologists, explained Gregory Stone, the New England Aquarium's Conservation Director and Co-leader of the expedition. Godzilla is the most prominent among any of a number of new icebergs this year by seawater that has become warmer in the Southern Ocean, a change that most scientists attribute to global warming. Many of the icebergs are off a larger piece of ice from the Antarctic ice shelf that extends from the frozen continent.

A glacier on the Monte Rosa Mountain near the triangular Matter Horn Peak, began melting in hot weather during the first of July 2002, and created a lake covering about 35 acres. Workers wanted to scoop up water and carry it by helicopter to another lake where it could drain down a stream. The operation was stopped by bad weather as it rained and visibility was almost zero. Then they resorted to pumps and hose to pump the water through.

Villagers below Monte Rosa in Macugnana were in a state of alert and ready to evacuate. The 100 million cubic feet of water could spurt out or seep through fissures in any direction at any time. The major worry was other glaciers on other peaks could start breaking away and melting also.

A 500 foot high glacier sent an avalanche of ice, rocks, and mud down a mountain side near the Russian village of Gizel leaving as many as 100 people missing and feared dead. Part of the village of Nizhny Karmadon was destroyed and another village where 50 people lived was nearly entirely covered in ice. Among those confirmed missing were 17 villagers, 14 hikers, and 40 people with a crew led by an actor directing a film. They were later found.

It slid down the Caucasus Mountains 15 miles pulling up trees and dislodging mud and rocks along the way stopping on the Gizel-Karmadon highway. It left a path about 300-400 yards wide.

A report issued by the United Nations Environment Program in April says at least 44 lakes in Nepal and Bhutan are filling so rapidly with icy water from melting glaciers that they could burst their banks within five to ten years. Hundreds of thousands of people could drown as roads, bridges, and hydroelectric plants wash away.

Geologist Jeffry Kargel of the U.S. Geological Survey in Flagstaff, Arizona, says the overwhelming majority of glaciers around the globe are retreating. Kargel is the International Coordinator of Global Land Ice Measurement from Space, a satellite program dedicated to photographing each glacier on Earth every year. He has no doubt about the cause. It's definitely due to global climatic changes, he says. Central and Southern Asia are most threatened by severe changes. Glaciers act as a liquid bank account for people in those areas, storing snow in winter and releasing melted water slowly in the summer. If glaciers disappear, summer crops will wither and die.

Sea level has risen between 12 and 20 inches along Maine's coast and as much as 2 feet in Nova Scotia during the past 250 years according to a team of international researchers. It is the biggest rise in the past millennium and global warming is to blame, said Roland Gehrels of the University of Plymouth in England.

Sea levels today are rising faster than at any time in the past when it was subject to natural climate change, a lead researcher said. He said, sea level rose at the end of the 18th century as a result of natural warming. The rate of increase slowed, but then increased in the 20th century as industrialization swept the globe. The findings were presented at the Geological Society of America's annual meeting last week in Boston and marked the first time sea-level changes have been dated beyond the time span of instrumental observations and verified at multiple locations.

Gehrels and Researchers from the University of Maine, the University of Plymouth, and Reading University in England studied sites at Machiasport, wells in Maine, and at Chezzetocook in Nova Scotia. Gehrels and his team reconstructed sea levels by using new dating techniques on salt marsh sediments. Gehrels drilled into the three locations and sampled sediments from the core. He was able to reconstruct the sea level by focusing on fossilized microorganisms called aminifera.

He determined how often the marshes were flooded by comparing the levels of foraminifera in the core with those typically found on the surface. That helped him to determine the sea levels. The researchers were able to determine sea level 1,200 years back at the Maine locations and 300 years back in Nova Scotia. Gehrels said his techniques worked particularly well in Maine because salt water marshes have not been disturbed by storms and erosion over the years. They are covered by ice for most of the year.

Sea level rose faster in eastern Maine and Nova Scotia because the coastline gradually sank as the last ice age receded north, Gehrels said. The sea level rise was 20 inches in Machias Port compared with 12 inches in Wells about 185 miles southwest.

Global temperatures could rise by up to 10.5 degrees F in the next century bringing increased floods, droughts and killer storms if people do not cut the air pollution that scientists believe is warming the atmosphere. The report from 99 nations meeting in China, joins a drumbeat of troubling news for people concerned about global warming.

From the setting string of 100 degree days in Dallas to the melting glaciers in the far north, bizarre climate events and early maple syrup season in New England have become water-cooler conversation nationwide.

The year 2000 was one of the warmest of the past 106 years, on record. The Unite States National Oceanic and Atmospheric Administration predicted that the average annual United States temperature for 2000 would be about 54.2 degrees F at years end.

At least 12 people in Alabama were killed when a band of the worst tornadoes in 70 years ripped through parts of the state. Scours of others were left injured as National Guard troops searched the twisted rubble for survivors. The tornadoes hit near the Alabama-Florida border, as well as on the outskirts of Tuscaloosa and in the rural area north of the state. One twister tore through a trailer park and an upscale community in Tuscaloosa, destroying houses and leaving mobile homes in mangled piles.

Fresh outbreaks of arctic air blew deep into the United States and Northern Mexico in one of the bitterest starts of winter in recent years. Snow fell as far south as Georgia and the Florida Panhandle, while areas farther north received significant amounts of ice and wind-driven snow. On the eve of the Winter Solstice, temperatures in Miami Beach plunged into the 40s.

Rampaging elephants devastated vast swaths of cropland and destroyed 44 communities in northern Nigeria since early December. Officials reported that the elephants has strayed from a nearby game reserve and recently stomped through the villages of Gwoza, Chibok, Damboa, Biu, Kwayaa-Kusar, and Bayo. One resident said "This is the worst elephant invasion of our area within the last 35 years. The elephants destroyed farmland, cash crops, and economic trees." The Agriculture Commissioner reported that the government was planning to fence in the Yukan game reserve to prevent the pachyderm from grazing outside its borders and to keep communities out of the path of animal annual migration routes.

An article in the National Geographic dated February 2001 said, sea levels are climbing and people are in their way. Coastal and Island dwellers risk losing their homes, or even their lives, as a consequence of rising oceans. In the past century sea levels climbed an average six inches. That figure seems too small to be of concern, but it reflects a rate of increase ten times the average over the past 2,000 years. The world is warming, causing seawater to expand and accelerating the melting of mountain glaciers. Melting polar ice sheets could eventually add to the rise.

Sea levels have fallen and risen over the millennia. At the peak of the Ice Age 20,000 years ago, average sea level was 400 feet lower. Even if the pollutants that contribute to temperature rise are reduced, the climb is expected to continue in the 21st century, raising sea levels along with it. The consequences will be, flooding, erosion, tainted drinking water, displaced population, and loss of farmland.

In the past century, a one degree Fahrenheit rise in global temperatures led to a six inch sea level rise. If the earth warms by four degrees by 2100, seas may rise another foot and a half inches. An average temperature from 57*F in 1900 to 58*F in 2000 caused the sea to rise 6 inches. An average temperature increase to 62*F in 2100 will cause the sea to rise 18.5 inches.

It is estimated that by the year 3000, the sea will rise 20 feet. It would cause a redrawing of the Earth's map. If global warming increase over this millennium and the ice melts, sea levels should creep up to 20 feet higher.

Populations concentrated in river deltas such as the Ganges in Bangladesh or the Nile in Egypt will be in immediate danger of losing their homes to rising sea levels, along with some of their countries richest farm land. A third of the people in the world lives in coastal zones, and by the end of this century low-lying islands in the Caribbean could be uninhabitable.

The Netherlands will spend more than $1 billion to build new dikes, bolster the natural sand dunes, and widen and deepen rivers enough to try and protect the country against a 3 foot rise in ocean levels.

Increased melting of the floating polar ice in the Arctic Ocean will raise sea levels. The danger lies in ice sheets attached to land masses like West Antarctica's coastal ice shelf. The rest of Antarctica is less vulnerable because even a rise in temperature of a few degrees would leave it below freezing and could lead to snow, which would simply add to the existing ice sheet.

Longest Drought from Asia to Africa

From Asia to Africa, severe, long-term drought now effects more than 20 countries. In August 2000, 100 million people were suffering from drought in Africa, Central Asia, parts of the Caribbean, and Central America. Floods wretched havoc in Cambodia, Laos, and Bangladesh. Now Mozambique is awash in floods, and a series of earthquakes has hit El Salvador. Mozambique's government in 2001 appealed for $30 million in aid.

In more tranquil Kenya, some 4.4 million people are estimated to be in urgent need of food assistance. In Nairobi, they graze their cattle in the parks of the capital city, so desperate to find grass for the cattle to eat. It is Africa and the troubled countries in the Horn of Africa where the long-term drought has had its most damaging impact, with some 16 million people facing food shortages according to UN statistics. Last year, a famine was only narrowly averted after a UN appeal.

Three years of inadequate rainfall in countries already reeling from poverty and civil unrest have reduced millions to desperate measures. In Afghanistan, tens of thousands of refugees have fled their homes in search of water and have eaten next year's seeds for food. In Tajikistan where 85 percent of the population lives below the poverty line, many of the men among the 1 million affected by drought are now migrating to Russia in hopes of finding food for their families. In war torn Angola, cattle have been dropping dead at the rate of 100 a day. In Ethiopia, the bleached bones of dead cattle liter the ground. In Sudan families are selling their precious livestock which is the equivalent of American's emptying their bank accounts for a handful to eat.

Because many of those countries are pastoral societies in which people rely on their livestock as their source of food, years without rain have had a particularly devastating impact. Camels, goats, and cattle have died or being sold off. The duration of drought for nearly three years with only intermittent rainfall has produced a surge in mortality, malnutrition, and infectious disease, said Dr. Peter

Salaam of the United States Centers for Disease Control in Atlanta. With people's immune systems weakened by hunger, infectious diseases like malaria and measles can rip through a population killing thousands.

In June 2001, hundreds of cholera cases were reported in drought areas. The disease also has been reported in parts of Ethiopia, where an estimated 6.2 million are thought to be affected by droughts. The World Health Organization has reported soaring rates of tuberculosis because of drought. In Sudan, where there is an 18 year old war between Islamic government to the North Christians and rebels in the South stopped the UN from carrying in food and aid in 1998. That probably produced 100,000 deaths in the South said Robert Winter, Executive Director of the United States Committee for Refugees in Washington, DC.

Combined with poverty and civil unrest, it is why a near-record 60 million people worldwide now desperately need emergency assistance including food and disaster relief specialists say. Drought and other natural disasters such as earthquakes and floods are now the leading causes of food emergencies. The number of people affected has soared from 4 million to 49 million since 1995 says the United Nations World Food Program. In the last four years, the numbers who are hungry, owing to drought, has more than quadrupled. Catherine Berating director of Rome-Based World Food Program said, there are so many countries affected.

Meteorologists say the droughts, alone with severe flooding in other parts of the world, may stem in part from erratic weather pattern by the La Nina phase or El-Nino, with a significant shift in the temperature of the Pacific Ocean off South America. Scientist in 2001 predicted that global warming could lead to growing numbers of similar natural disasters that would have their most direct effects on the world's poorest countries.

Harvard University professor James McCarthy, chairman of the International Panel On Climate Change said, we see clear indications that there will be more extreme events. The countries that will experience the most devastating effects of this are the least equipped to handle it.

People are not just hungry, they are without any future means of livelihood. That makes it important to provide drought stricken areas with medicine and food for livestock. Tufts University Lautze said, he was very worried about the long-term future of these societies, and the world community is getting very tired of "responding."

No Seal Pups on Ice

Rick Smith, is a Marine Biologists and Canadian Director of The Anti-Sealing Group. An early melting hurts seal hunters in Canada. Campaigned to save the seal, they believe this year's near-absence of ice floes in birthing regions of the gulf may cause a catastrophic loss of newborns. Harp seals give birth on the floes, and pups need days on the ice before they complete nursing and can take to the sea on their own.

At that time of year, ice packs normally stretch from Quebec's Magdalene Islands south to Prince Edward Island, and by late March, the floes should be teeming with hundreds of thousands of seal mothers.

"In five days of flying over the entire region, we were not able to spot a single seal pup," Rick Smith said in an interview from Prince Edward Island. Normally, there are 200,000 to 300,000 harp seals born in the Gulf of St. Lawrence. Smith added, "this could spell devastation for the population, not only in the Gulf, but off Newfoundland, where the hunt may become even more intense to compensate for the lack of seals in the Gulf of St Lawrence."

"The seals need ice, but whether there has been a real reproductive failure this year remains to be seen," said Ian Mclaren, professor of biology of Nova Scotia's Dalkhotisie University and an authority on seals. One year's loss of pups is not necessarily a catastrophe. The Canadian Cull is the largest hunt of seagoing mammal in the world, and has been the target of a global crusade since 1977.

What they found this year was perhaps more horrifying than blood-drenched ice and mother seals wailing for their slaughtered young, Smith said.

Sealing vessels from the Magdalene Islands, usually out in force in late March, are tied in port. Although the hunting quota for the gulf was set at 77,000 animals, the season is finished with only a few hundred seals taken. The real fear is that hundreds of thousands were stillborn or drowned at birth.

Seals, which are clubbed to death or shot with high-powered rifles, are most valued for their find pelts, although demand in Asia for powered male seal sex organs, considerable an aphrodisiac, has created a new market in recent years. Seal oil, rich in Omega-3 fatty acid that may be helpful in reducing blood cholesterol levels, is the other main product.

This year, for the first time in 30 years, there will be none of the ritualized clashes between sealers, and seal lovers on the ice of the gulf because the ice is gone. Smith and other anti-sealing campaigners packed their gear late last week and headed home.

Some scientists think many of the seal populations have declined over the past years because they were trapped in a net, rope, line and drowned. Whales and dolphins also get caught in old fishing nets. Ocean garbage can be a serious threat to seabirds and shorebirds. They also become caught in old fishing nets, fishing line, or plastic beer and soda container rings, and drown or choke to death. Pelicans get the fishing line of plastic beer container rings caught around their beaks so they cannot catch their food thus they starve. Pelicans and egrets get the fishing lines caught around their wings and legs so they cannot fly, and some birds and sea turtles eat the little pieces of plastic, thinking it is food and die a miserable death.

Disease Threat Because of Warming

As the warming temperatures increase around the world, there is a trend that could mean more epidemics in humans, plants, and animals according to a report in a Science Magazine. Oyster disease is in Maine waters, the Dengue Virus is in Latin America, and Rift Valley fever in the Middle East. Rift Valley causes people to vomit blood. It spread across the Red Sea in 2000 killing 200 people in Yemen and Saudi Arabia.

New England is losing cold weather that is a good defense of diseases. Normally, every fall mosquitoes that may carry the deadly West Nile virus are killed before they multiply and spread the disease widely. A small rise in global warming disease-carrying organisms may regenerate faster or go into new areas where populations may have little or no natural resistance. Rick Ostfeld an animal ecosystems person say's, their population could double with only a half or with a single degree of warming. The report also says that with increased temperatures, mosquitoes that carry the Dengue Virus bite more often, parasites that attach to butterflies gather in greater density, and slime grows faster.

Many environmentalists also blame pollution for the rising number of fish kills, whale kills and red tides that are becoming familiar events in coastal communities from Maine to California.

Despite improvements in water clarity over the last two decades, American's coastal waters, including the Great Lakes, continue to have serious pollution and ecological problems according to an Environmental Protection Agency study released recently. The study found that while 56 percent of American coastal waters are considered clean enough to support plant and animal life and for human uses, about 34 percent are in too poor condition to support aquatic life and 33 percent are considered unacceptable for human use.

The report comes as scientists continue to puzzle over a 100 mile-long, oxygen-eating "black blob" possibly made up of microscopic algae that mysteriously appeared off Key West, Florida, last month and is now dissipating. A similar but unrelated "dead zone" off the mouth of the Mississippi river has been a source of worry for several years.

The new EPA study found water clarity good in the Great Lakes, the Northeast coastal waters running from Virginia to Maine, and along the West Coast, but only fair in the Gulf of Mexico and along the southeastern shoreline.

Prospects for The Future

An Intergovernmental Panel on Climate Change met in Shamhai and officially released a report declaring that global warming is not only real but man-made. The decade of the 90s was the warmest on record, and most of the rise was likely caused by the burning of oil, coal, and other fuels that release carbon dioxide, as well as other greenhouse gases. Future changes will be twice as severe as predicted just five years ago. Over the next 100 years, temperatures are projected to rise by 2.5 to 10.4 degrees worldwide, enough to spark floods, epidemics, and millions of environmental refugees.

The hotels that now line Miami's South Beach could stand waterlogged and abandoned. Malaria could be a public health threat in Vermont. Nebraska farmers could abandon their fields for lack of water. Outside the United States the impact would be much more severe.

Rising sea levels could contaminate the aquifers that supply drinking water for Caribbean Islands, while entire Pacific island nations could completely disappear under the sea. Perhaps the hardest hit country would be Bangladesh, where thousands of people already die from floods each year. Increased snowmelt in the Himalayas could combine with rising seas to make at least 10 percent of the country uninhabitable. The water level of most of Africa's largest rivers, including the Nile, could plunge, triggering widespread crop failure and idling hydroelectric plants.

The developing world will be hit hardest and least able to cope. Bangladesh will be hurt, says Stephen Schneider, a climatologist at Stanford University noting that flooding there and in Southeast Asia and China, could dislocate millions of people.

While governments argue over what is to be done, major corporations such as BP, Amoco, and DuPont in America are retooling operations to reduce greenhouse gases. The largest coal-burning utility company is experimenting. American electric Power of Columbus, Ohio, is testing "Carbon Capture." Which would separate carbon dioxide emissions and dispose of them in deep underground saline aquifers effectively causing carbon emissions free coal power. Over the next decade, the Netherlands will spend more than $1 billion to build new dikes, bolster the natural sand dunes, and widen and deepen rivers enough to protect the country against a 3-foot rise in ocean levels.

The effects would vary wildly from one place to the next; what might be warmer winters in Fairbanks, Alaska would be bad for another. Weather would become more unpredictable and violent, with thunderstorms sparking increased tornadoes and fighting a major cause of forest fires. The effects of El Nino, the atmospheric oscillation that causes flooding and mud slides in California and the tropics would become more severe. Natural disasters already has high cost. In the 1990s, the tab was $608 billion, more than the four previous decades combined, according to World Watch Institute.

Deaths would also increase from natural disasters, and warmer weather would affect transmission of insect-borne diseases such as malaria and West Nile Virus, which made a surprise arrival in the United States in 1999. We do not know exactly how West Nile was introduced to the United States, but we do know that drought, warm winter, heat waves, and the conditions that helped amplify it says Paul Epstein, a researchers at Harvard's School of Public Health.

Andrew D. Anderson

Rain would become more frequent and intense in the Northern Hemisphere. Snow would melt faster and earlier in the Rockies and the Himalayas, exacerbating spring flooding and leaving summers drier. Sea level worldwide has risen 9 inches in the last century, and 46 million people live at risk of flooding due to storm surges. That figure would double if oceans rise 20 inches.

The IPCC predicts the seas will rise anywhere from 3.5 to 34.6` inches by 2010, largely because of thermal expansion. Warmer water takes up more space, but also because of melting glaciers and ice caps. A 3 foot rise at the top range of the forecast, would swamp parts of major cities and islands, including the Marshall Island, in the South Pacific, and the Florida Keys.

By 2015, 3 billion people will be living in an area without enough water. The already water-starved Middle East could become the center of conflicts, even war over water access. Turkey has already diverted water from the Tigris and Euphrates rivers with dams and irrigation systems, leaving downstream countries like Iraq and Syria complaining about low river levels. By 2050, such down stream nations could be left without enough water for drinking and irrigation.

Projected changes:

1. Australia: The Great Barrier Reef could be ruined as a tourist attraction if the water temperatures increases by 3.6 degrees.
2. Bangladesh: Faster melting snowcaps in the Himalayas, rising sea levels, and cholera outbreaks could force millions from their homes.
3. Brazil: Models project that populous Northeastern Brazil could suffer some of the most severe crop setbacks because of a drought.
4. Coral bleaching: Warmer water could bleach coral reefs, leading to their destruction. This may deplete fisheries, disrupting food supplies and tourism.
5. Crops: Drought and high temperatures could cause crop failure and malnutrition.

196

6.   Disease: Warmer, wetter conditions may amplify insect-borne diseases, such as malaria. Flooding could spawn more water-borne illness.
7.   Fires: Drier summers and higher temperatures create ideal conditions for wildfires. In 1997, some 40,000 people were treated for smoke inhalation in Southeast Asia.
8.   Floods: Sea levels will rise in the next century, leaving people more vulnerable to storm surges. Earlier melting snow could cause rivers to overflow.
9.   Marshall Islands, Tuvalu, Kiribati: Swelling oceans could cover these islands, forcing residents to evacuate.
10.   Mexico: Rising temperatures could cut maize crops by 20 to 60 percent.
11.   Nigeria: A 3 foot rise in seal level could displace 4 million people and leave parts of the capital city, Lagos, underwater.
12.   Pollution: Sunlight breaks pollution into noxious substances, causing more respiratory problems.
13.   South Africa: Malaria may surge in areas previously too cold for mosquitoes to inhabit.

A Dismal Future

Peter Gleick, president of the Pacific Institute for Studies in Development, Environment, and Security in Oakland, California says, no matter what we do to reduce green house-gas emissions, we will not be able to avoid some impacts of climate change. The newest global-warming forecast is backed by data from myriad satellites, weather balloons, ships at seas, and weather stations. Large computer models of the global climate system also gives data.

A tropical island paradise and glistering Alpine skiing retreats may be lost in future generations, while melting ice caps in polar regions could unleash climate changes that could continue for centuries, according to a UN Report released in 2001. The melting of

Equatorial Glaciers in Africa and Peru is another powerful indication of global warming. Ice tops on African's Mount Kilimanjaro, and others in Peru and Tibet, may be disappearing the victims of a process of shrinking mountains glaciers everywhere.

Releases of carbon dioxide and other so-called greenhouses gases could trigger an abrupt and dramatic change in global temperatures that governments are unprepared to cope with, according to a report from the National Research Council. They suggest that the current rapid rise of carbon dioxide in the atmosphere, due mainly to burning fuels such as oil and coal, could trigger a new round of rapid heating or cooling. Authors argue, greenhouse gases could quietly build up in the atmosphere until they cause a sudden shift in ocean circulation patterns, bringing heat to some places and a new cold to others.

When working normally, the circulation cycle brings warm water up from the tropics where it merges with the eastward moving Gulf Stream and travels up the coast of Europe. Easterly winds warm up as they blow across the current, keeping some European cities relatively warm.

Global warming could disrupt this cycle by melting glaciers and increasing overall evaporation, which would lead to increased precipitation, changing the density of northern water and stopping the circulation patterns. That would deliver sudden cold to places such as Western Europe, and increased heat elsewhere.

In Santee Fe, NM, the rose bushes will go without water because every drop of moisture will be spent trying to save the large evergreen trees that shade the low-slung building from the blazing sun as the drought grips the West. From central Montana through Wyoming, Colorado, Utah, New Mexico, Arizona and parts of Texas, rivers are running low, reservoir levels are dropping, forests fires raging, and cities are implementing emergency measures, all before the dog days of summer arrive.

Officials are moving to free up state emergency funds and get low-cost federal loans to farmers and ranchers. With extreme dryness exacerbating tinder box conditions in many forests, the West's fire season started preparing almost two months early this year said Michael Hayes, climate impact specialist with the National Drought Mitigation Center in Lincoln, NB. Snow accounts for as much as 75 percent of water available in this part of the country, but this year, some mountains tops are already showing their gray peaks through a thin snow pack.

The Rio Grande which is running 8 to 20 percent of its normal levels in parts of southern Colorado and New Mexico has been particularly hard it by the lack of mountain runoff.

Texas will have to stop the urban sprawl that increases smog by inducing people not to drive their vehicles as much. The Texas Department of Transportation forecast another 45 percent increase in driving statewide by 2017, on top of a 325 percent increase since 1967. It will not be able to stop the urban areas from growing as more people move there and more houses are built. It is necessary that a commuter-rail be built for less pollution and more conveniences.

Most of the world's energy comes from oil, coal, and natural gas that emits some 22 billion tons of carbon dioxide into the Earth's Atmosphere each year. Such emission could rise 55 percent by 2020 as populations swell, according to the United States Department of Energy.

Coal is the filthiest fuel gouged from the ground. It has more carbon than any other fuel. When burned, it releases carbon, mercury, lead, and sulfur into the air. China and India are projected to account for the greatest rise in the use of coal. With only 7 percent of the world's population, North America consumes 30 percent of the world's energy. Such gluttony could shift as developing nations, mainly in Asia and South America demand more.

It is imperative to accept the idea that we are connected to all things in this world and when there is a change in one area we all are effected.

There are other areas that the result of warming is harmful to our health that most people are not even aware of. Worldwide warming is blamed on the decline of coral reef in many parts of the world. Bleaching occurs when coral is diseased or under stress and expels the symbiotic that it normally harbors. The calcium carbonate skeleton then shows through, giving a pale or stark white appearance to the coral.

Chapter 12: Coastal Waters

Waters We Swim In

The New York problem is not unique, along with a gruesome array of other garbage. Hospital waste has floated ashore on beaches from Maine to Texas in the past year. Beach closing because of bacterial contamination is even more common. This summer, public health authorities are in locations because of sewage plant breakdowns. Nearly a third of Louisiana's oyster beds are routinely closed because of pollution, and half the shell fish beds in Galveston Bay, Texas, are off limits to fishermen. Temporary or not, all these local pollution alerts underscore what responsible environmentalists say is not a full blown national crisis, it is the wholesale contamination of the United States Coastal areas by billions of tones of sewage, garbage, toxic chemicals, and other contaminants. Although the United States is now disposing of its human and industrial waste more carefully than ever before, there is mounting evidence that the ocean, like the land, is faring badly under the ecological stress imposed by man.

As part of the Water Quality Act of 1987, the United States government is working to improve the condition of the nation's beaches and coastal waters through the National Estuary Program. An estuary is an inlet, bay or an area where a river meets the ocean. Estuaries and the nearby coastal areas are among the richest and most useful areas in the world. Fish and shellfish lay their eggs and grow up in estuaries, and about two-thirds of all fish caught are hatched in estuaries. Birds and many animals live in the wetlands that border them.

An estimated 75 percent of the population of the United States lives within 50 miles of the coastline. That means major cities with sewage plants, industry with its industrial wastes, and millions of tourists pour their garbage into these estuaries. Because living near an estuary is so attractive, more and more people are moving to these areas and producing more and more garbage. The land is rich so farmers farm the land. The fertilizers and pesticides they use often run off the farmers land into streams and rivers and then into the estuaries, polluting the water and threatening sea life.

The National Water Quality Inventory in 1994, reported the states had assessed three-fourths of the 36,890 miles of estuaries located along the coastline of the United States running from Puget Sound in Washington State to Galveston Bay, Texas, and to Buzzards Bay, Massachusetts. The biggest problem faced by estuaries is development, building of houses, roads, malls, factories, and schools. With more people, more businesses, and more industries come added pollution. States, not federal government, have to make the big decisions on the type of businesses and housing that can be rebuilt and where they can be built.

Fish contamination was at a high level in the Great Lakes, the Northeast, and the Gulf of Mexico, while the West was rated, "fair," and the Southeast "good," in this area of concern. Of 1,444 coastal beaches examined in the survey, 370, or 26 percent had been closed or issued a contamination advisory at least once in 1999, the report said. Reasons for the closures included sewage and elevated bacterial level resulting from storm water runoff, pipeline breaks, sewer overflows, and unknown causes.

The problem for most landlubbers is that most of the effects of coastal pollution are hard to see in bays and estuaries that are now in jeopardy. Boston Harbor or even San Francisco Bay is still delightful to look at from shore. What is happening underwater is quite another matter. Scuba divers talk of swimming through clouds of toilet paper and half-dissolved feces, of bay bottom covered by toxic combination of sediment, sewage and petrochemical waste appropriately known as "Black Mayonnaise." Fishermen haul in lobsters and crabs covered with mysterious burn holes and fish whose fins are rotting off. Offshore marine biologists track massive tides of algae blooms fed by nitrate and phosphate pollution-colonies of floating micro-organisms that strangle fish by stripping the water of its life-giving oxygen.

Skin-divers and scuba divers have been caught by old nets and fishing lines, they wrap around boat propellers, and shellfish that live in polluted waters can become very dangerous to eat and swimming in polluted waters make people sick.

Storm sewers connect the numberless drains on the streets of almost every city and town in America that dispose of rainwater. It is a fact of urban life that storm sewers also collect large amounts of pollutants: gas, oil, antifreeze, fertilizer, and pesticides, then deposit the resulting mess in the nearest waterways. Christine Reid, in Santa Monica, California, blames Los Angeles area storm sewers for much of the pollution in Santa Monica bay. Many of us just aren't attentive to things we do not see. The people sweeping poo in the drain do not realize that their kids will be surfing in it.

Medical wastes called attention that many by products of modern technology are not easily disposed of. Despite the fact that many contamination problems are much more serious, the sensational nature of medical wastes prompted quick legislation. A public outcry prompted Congress to pass the 1988 Medical Waste Tracking Act requiring producers of medical waste to be held accountable for safe disposal or face up to $1 million dollar fines and five years in prison.

During the summers of 1987 and 1988, medical wastes from hospitals, including containers of blood and syringes used to give shots, floated up onto the shore along the East Coasts in New York and New Jersey. Some were infected with HIV and hepatitis. The government closed the beaches in those areas for several weeks, but medical trash washed up on shores along the Great Lakes and those beaches had to be shut down also.

One of the greatest solid waste threats to marine life is the plastic products that are dumped daily along with the rest of the garbage by ocean going vessels, commercial and recreational fishing boats, offshore oil and gas platforms, and military ships. As discarded trash drift toward land, plastic bags and pellets are ingested by fish, turtles, birds, and marine mammals. Other types of plastic debris come from land sources. These include factory waste, overflows, illegal garbage dumping, and human littering. Dead whales and dolphins have been found with their stomachs full of plastic bags. Not only do adult birds eat small plastic pellets resembling fish eggs and feed them to their young, they often become trapped in the plastic carriers used for six packs. An estimated two million sea birds and 100,000 marine animals die each year as a result of ingesting or becoming entangled in plastic.

Commercial fishing nets are now made of durable, non-degradable plastic, but when they are lost or discarded in the water, they pose a particular hazard to seals, dolphins, whales, and diving birds, which can become entangled in the nets. One scientist reported that 56 percent of endangered whales for the New England Aquarium had scars from plastic gill nets or lobster gear entanglement. In the 1987 cleanup of the beaches in Texas, the garbage collected was found to contain large amounts of plastics.

Despite the redevelopment of the Boston waterfront, Boston Harbor was one of the worst coastal cesspools in the nation. Three million people live within a twenty five (25) mile radius of the harbor, and nearly all the human and industrial waste of the metropolitan area wound up in its waters. The harbor bottom contains high concentration of DDT and PCB, a chemical compound that was widely used in electrical equipment and suspected of being a carcinogen. Local beaches are littered with grease balls, tampon applicators, and condoms, all of which are apparently released by antiquated and overloaded sewage systems.

The Boston system offers what is politely known as "primary" treatment which means that sewer sludge is separated from waste water, and then both are dumped into the harbor. If that sounds unbelievable, it is. After a heavy rain storm the whole system breaks down and raw sewage, plus whatever is flushed off local streets, is channeled into the harbor through combined sewer overflows. Boston has cleaned up its act, but only after local environmental groups took the local sewer authority to court. The authority stopped dumping sewage sludge into Boston harbor by 1991, and the beaches were free of garbage by 1995.

Sand Problems and Erosion on Beaches

Man with his expertise and brain thinks he has a way to stop erosion on beaches. He with his just plain common cense, should know that the ocean is an awesome force, constantly washing sand and lots of other things from one place to another. But that has not stopped coastal residents from trying to fight it, erecting walls, groins, and more exotic defenses against the tides. Nothing can stop the sea, and only the costly process of pumping new sand onto the beach seems to slow erosion in an environmentally acceptable way.

Massachusetts is considering a large-scale public program to replace sand on the eroded beaches that had cost other coastal states hundreds of millions of dollars. Sand is pumped from the sea floor onto a barge, then pump through a temporary pipeline to the beach where bulldozers fashion new sand dunes. Beach renourishment provides a temporary buffer against the sea, but erosion eventually carries the sand away.

Included in this process, bulldozers scrape sand from the beach and push it toward the dunes, creating a buffer that protects property owners. However, bulldozing may actually stop erosion for a short period of time by creating steep-angled beaches that get hit hard by waves.

Beaches continually lose and collect sand, sometimes eroding for decades before entering years-long periods of growth. Most of the sand came from the erosion of coastal bluffs, where retreating glaciers deposited tons of sand and gravel, 10,000 years ago. The ebb and flow of the shoreline is perfectly natural, but becomes a problem when buildings are to near the water.

Why Seawalls Collapse

A seawall is usually built vertically from the shore to protect the shore from erosion. When a wave of water hits the wall, it deflects the energy of the wave instead of allowing it to dissipate on to the surface of beaches. The reflected energy creates turbulence and the wave is directed downward which erodes sand at the base of the sea wall. Eventually, waves sweep under the base, causing the wall to tip over.

Groins are built to prevent sand from shifting sideways down the beach. A groin is a rigid structure built out from a shore to protect the shore from erosion or to trap sand. They interrupt the natural southward drift of sand. Sand collects on the north side of the groin, but the barrier prevents sand from reaching beaches father south. Over time, the property owner on the north side gets more sand at the expense of his neighbor.

Waves take away a fine layer of the beach every time they slap the shore, but the sand doesn't simply go out to sea. In New England, waves tend to approach from the north creating a current along the shore that carries sand slowly to the south.

From Plum Island to Nantucket, coastal Massachusetts is losing against the sea. Neither wall nor other barriers have stopped waves from carrying away chunks of beach and dunes. At the current rate, natural erosion and rising sea level could destroy one out of four beach front homes over the next 60 years, according to the Federal Emergency Management Agency.

Massachusetts is cracking down on the cheapest defense using bulldozers to build dunes. Much to the dismay of their neighbors, Department of Environmental Protection Officials rejected their plan to bulldoze the beach, saying that strategy actually speeds erosion by feeding more land to the sea. "We know that pushing sand from the beach onto the dunes is a temporary solution, but sometimes a temporary solution works," said a neighbor at a recent meeting with state officials. Maybe none of us should live here, but we do."

Massachusetts is considering a new potentially costly approach to protecting the coast. Dredging sand from offshore and delivering it to erosion hot spots. States such as Florida and New Jersey have spent hundreds of millions of dollars pumping sand ashore, a process called "beach re-nourishment." The strategy is strictly temporary, but, unlike the bulldozer approach, re-nourishment brings new sand to the beach and its dune system. But there is a high cost for a temporary half of a solution.

No one has been able to come up with an affordable, environmentally sound way of protecting waterfront property form erosion. And many of the things people have been doing for years to protect their homes have been making matters worse. Beaches erode away in front of sea walls. Groings, the long arms of rock that jut out from so many beaches, stop erosion on one beach by causing it on the next. On some beaches, bulldozing or beach scraping is seen as a compromise. On others, it offers short term protection at the expense of the long-term life of the beach.

People are really more concerned with their immediate needs and it is hard to get them to think about a long-term beach wide approach, said Thomas Skinner, director of the Office of Coastal Zone Management. It is the timing we are concerned about, Tolpin, an ophthalmologist, told Coastal Zone management officials. We are going to do the right thing, but we just want to get the sand up to where we are protected.

There is another drawback to re-nourishment, price. With no state funding program yet, it is going to cost a lot more than the few hundred dollars they each chip in for a bulldozer. It will cost $4.5 million for 10 years, according to a study by the Army Corps of Engineers, which coordinates beach rebuilding projects for the federal government. Even if the state plan becomes a reality, re-nourishment is less than a perfect solution. Some communities have watched millions of dollars in sand wash away with the first big storm, while other re-nourished beaches may last 10 years or more. And proposed budget cuts mean Massachusetts may not be able to rely on the federal funding that has paid for many of multi-million-dollars beach projects.

A study by James O'Connell, a coastal process specialist for the Sea Grant Research Program in Woods Hole, found that many Cape Cod towns are still approving waterfront projects that put homes at risk and contribute to beach erosion. O'Connell counted 11 new homes and 22 additions allowed on dunes and erosion-sensitive barrier islands in the 15 Cape towns in 1999.

When the Ice Age glaciers that scraped across North America melted away, they left the future beach lovers of Massachusetts a gift, silt, clay, and sand from the glacial cliffs along the coast much of the sand for the state's beaches. At the Cape Cod National Seashore the cliffs crumble constantly, replacing sand that is washed down shore or out to sea. Since there are no homes at the top, the cliff erosion is not a big problem. Whatever sand that once fed the beaches, there is none locked up behind the sea walls and whatever beaches were once there are gone or shrinking.

On some beaches, homeowners have rebuilt the dunes by putting up temporary fences to collect sand and then reinforcing it by planting dune grass. If the dunes grow larger, they catch sand, feed the beach, and protect the homes behind them. But they can also be quickly wiped out in a storm and they were on Plum Island in March 2002.

State biologist Rebecca Haney says that it is not uncommon for beach residents to claim erosion is not a problem. According to the latest data, which compared the 1978 shoreline with the 1994 shoreline, Humarock Beach is eroding about two feet a year, she said. The state will press ahead with plans for a beach re-nourishment program, but even that is not a good strategy for areas with high erosion rates. The state would also like to try again to set up a fund to buy out homeowners who are tired of fighting.

Chapter 13: Genetic Help for The Future

Vulnerability to Food-Borne Diseases

Safe food is a moving target because we are always moving. Our eating habits and our ways of producing food change. In this country the number of people most vulnerable to food borne disease is growing. Within the next three decades a fifth of the population will be over 65, and many of them will be particularly susceptible to serious infection from Salmonella, Listera, and E. coli. Young children are more likely to be exposed to these diseases than they were a generation ago, not because the production of food has changed, but because families eat out or take home prepared food more often.

These microbes themselves are changing in new populations through new food vehicles, and causing more or new disease. We still have limited understanding of how food borne pathogens work. After nearly 20 years of research, we still cannot consistently treat advanced E. coli infections. We still are searching for clues to how food pathogens spread among cattle, egglayng hens, and broiler chickens.

E. coli in Premature Babies

The year 2002 shows E. coli is rising in preemies and antibiotic is blamed in the spread of bacteria. The rise among premature babies has overtaken strap as the most common infection in infants. Scientist do not know what this series is because E. coli bacteria can be more deadly than streptococcus germs. Streptococcus blood infections in new born preemies fell to nearly three-quarters during the 1990s probably because more women in labor received antibiotics to keep from passing on the bacteria to their babies the researchers said.

During the same period, E. coli infections doubled and it is believed because ampicillin, the antibiotic commonly used to wipe out strap, gave E. coli time to flourish. The study was led by Dr. Barbara J. Stoll, Professor of Pediatrics at Emory University School of Medicine. E. coli is among the bacteria that live harmlessly in many people's intestinal tract. If they spread into a woman's vagina during pregnancy, they can overwhelm the newborn's weak immune system, sometimes causing mental retardation, hearing, vision loss or death. The researchers studied records on 5,144 babies born at about 170 hospitals around the country.

Some of the questions we need to answer, how can we dispose of animal manure without threatening the environment and the food supply? How can we make safer the food and water that animals consume? How can we ensure the safety of imported foods and foods handled in our restaurants and kitchens to minimize our risk of infection from food borne disease?

Where Does The Diseases Come From?

Every year 74,000 Americans become ill and children die from E. coli. Most of the people in the United States do not realize they are and have been eating genetically engineered foods since the mid-1990s. More than 60 percent of all processed foods on United States supermarket shelves, including chips, pizza, ice cream, cookies, salad dressings, corn syrup, and baking powder contain ingredients from engineered soybeans, corn, or canola.

In the past decade, the biotech plants that went into house processed foods have leaped from hothouse oddities to crops planted on a massive scale on 130 million acres in 13 countries, among them Argentina, Canada, China, South Africa, Australia, Germany, and Spain. On United States farmlands, acreage planted with genetically engineered crops jumped nearly 25 fold from 3.6 million acres in 1996 to 88.2 million acres in 2001. More than 50 different designer crops have passed through a federal review process, and about a hundred more are undergoing field trials.

We are really hearing about genetic modification now, but the smart humans have been altering the genetic makeup of plants for years. Keeping seeds from the best crops and planting them in following years, breeding and crossbreeding varieties to make them taste sweeter, grow bigger, and last longer. In this way we have transformed the wild tomato from a fruit the size of a marble to today's giant juicy beefsteaks. From a weedy plant called teosinte with an ear barely an inch long has become our foot-long ears of sweet white and yellow corn. In just the past few decades, plant breeders have used traditional techniques to produce varieties of wheat and rice plants with higher grain yields.

Man continues to due research over time. In the past they crossed organisms that genetic make ups were similar thus transferring tens of thousands of genes. During these advanced years, genetic engineers could transfer just a few genes at a time between species that were distantly related or not related at all.

They can pull a desired gene from virtually any living organism and insert it into virtually any other organism. They can put a rat gene into lettuce to make a plant that produces vitamin C or splice genes from the cecropia moth into eggplants, offering protection from fire blight, a bacterial disease that damages apples and pears. The reason is to insert a gene or genes from a donor organism carrying a desired trait into an organism that does not have the trait.

Transgenic is the engineered organisms scientists produce by transferring genes between species. Some of the transgenic food crops on the market are varieties of corn, squash, canola, soybeans, and cotton from which cotton seed oil is produced. Most of theses crops are engineered to help farmers deal with age-old agriculture problems as weeds, insects, and disease, but will it harm humans?

Farmers spray herbicides to kill weeds, but biotech crops can carry special tolerance genes that help them withstand the spraying of chemicals that will harm nearly every other kind of plant. Some biotech varieties make their own insecticide, thanks to a gene borrowed from common soil bacterium or Bt for short. Bt genes are a code for toxins considered to be harmless to humans, but lethal to certain insects, including the European corn borer and insects that tunnels into cornstalks and ears making it good for corn farmers. So effective is Bt that organic farmers have used it as a natural insecticide for decades. When corn borer caterpillars bite into the leaves, stems, or kernels of a Bt corn plant, the toxin attacks their digestive tracts, and they die within a few days.

Squash and papaya have been genetically engineered to resist diseases and scientists have been experimenting with potatoes, modifying them with genes of bees, and moths to protect the crops from potato blight fungus, and grapevines with silkworm genes to make the vines resistant to Pierce's disease, spread by insects.

Scientists have gone farther by creating transgenic animals. Atlantic salmon grow more slowly during the winter, but engineered salmon, "supped-up with modified growth-hormone genes from other fish," reach market size in about half the normal time in a much larger size. It is given a modified gene that let it grow at a faster pace, while its counterpart grows slower in the winter. They are also using biotechnology to insert genes into cows and sheep so that the animals will produce pharmaceuticals in their milk. Will any of these changes cause sickness?

A fish that grows that fast could revolutionize the industry, allowing farmers to harvest two crops in the time it takes one to grow. Aqua Bounty Farms in Massachusetts had developed a transgenic Atlantic salmon. It has the growth hormone of the Chinook salmon and the promoter sequence, a molecular switch that turns on growth-hormone production of the ocean trout, a fish that grows year-round rather than just during warm weather months. The question for the FDA, researchers, and environmentalists is about risk assessment.

Aquaculture fish are kept in net pens in the ocean, but containment is not 100 percent, some of them will escape. What would happen if growth-enhanced Atlantic salmon got loose? Would the transgene propagate itself and potentially replace the wild population? Would the modified fish out compete its wild brethren? The potential harm, says Purdue University geneticist William Muir, is that transgenic fish may alter the niche or adversely affect other species.

Joseph McGonigle of Aqua Bounty, says the fish they sell will be female and sterilized so they cannot reproduce. Devlin says in our experience, we cannot achieve 100 percent sterility. Experiments show that some transgenic fish do out compete well with wild fish for food or mates, but no one knows whether that behavior will translate into greater fitness in a natural setting.

Is It Safe For Humans?

In the mid-1990s a biotech company inserted a gene from the Brazil nut into a soybean. The Brazil nut gene selected made a protein rich in one essential amino acid. The aim was to create a more nutritious soybean for use in animal feed. Because the Brazil nut is known to contain an allergen, the company also tested the product for human reactions, with the thought that the transgenic soybean might accidentally enter the human food supply. When test showed the humans would react to the modified soybeans, the project was abandoned.

For some consumer groups and scientists it raised the specter of allergens or other hazards that might slip through the safety net. Scientists know that some proteins, such as the one in the Brazil nut, can cause allergic reactions in humans, and they know how to test for the allergenic proteins. But a possibility exists that a novel protein with allergenic properties might in a new food produced by conventional means go undetected. Critics also say, the technique of moving genes across dramatically different species increases the likelihood of something going bad. The function of the inserted gene or in the functions of the host DNA, raise the possibility of unanticipated health effects.

Andrew D. Anderson

An allergy scare in 2000 centered around Star Link, a variety of genetically engineered corn approved by the United States government only for animals use because it showed some suspicious qualities, among them a tendency to break down slowly during digestion, a known characteristic of allergens. When Star Link found its way into taco shells, corn chips, and other foods, massive and costly recalls were launched to try to remove the corn from the food supply. In events like this, massive recalls are made, but nothing is said nor can much be said about the humans that got sick or will get sick in the future. All new foods present new risks and can rigorous testing minimize those risk, not "stop them?"

There maybe some health benefits. Genetically engineered for insect resistance may enhance safety for human and animal consumption. Corn damaged by insects often contains high levels of fumonisins, toxins made by fungi that are carried on the backs of insects and that grow in the wounds of the damaged corn.

How Are Genes Altered?

Scientists continue to find new ways to insert genes for specific traits into animal DNA and plants changing the food we eat. Dean Della Paenna envisions tomatoes and broccoli bursting with cancer-fighting chemicals and vitamin-enhanced crops of rice, and sweet potatoes for food for the poor. He envision wheat, soy, and bananas that deliver vaccines, peanuts free of allergens, and vegetable oils loaded with therapeutic ingredients that doctors prescribe for patients at risk for cancer and heart disease. He is a Plant Biochemist at Michigan State University and believes genetically engineered foods are the key to the next wave of advances in agriculture and health.

In North Europe and North America, the value of genetically engineered food has become subjects of many discussions. A researcher, Ajay Garg of Cornell University, planted and examined rice plants engineered to tolerate conditions that would kill ordinary rice. Inserting a pair of bacterium genes into rice DNA produced trehalose, a sugar that kept cells from disintegrating under stresses such as drought, salt and cold, which limit rice production worldwide. Since about 30 percent of the world's calories come from rice, this could help feed a growing population.

Ted Thannhauser, director of Cornell University's Bioresource Center says, using a machine called a DNA synthesizer, a technician at the center assembles different combinations of the four nucleotides, abbreviated as A, C, T, and G, that make up a gene. The order of the combination determines the traits that the gene will express.

One method of introducing selected genes into the chromosomes of plants involves inserting the gene into a soil bacterium called *Agrobacterium,* which is then introduced to the target organism, bringing along part of its own DNA as well as the added gene.

A more common method is a device known as a gene gun which was invented by a scientist who started out using a BB gun prototype to shoot genes into living cells. Now researchers at Cornell and elsewhere use a more sophisticated version to propel tiny particles of gene-coated tungsten or gold directly into cells. Completed, scientists watch for cell growth to see which cells have successfully incorporated the new genes. From ten rice plants, Cornell scientists modified only one or two that will be suited for further development.

The researchers monitored a large number of specimens in dishes to determine which plants will offer improvements over the original rice. The different between a successful and unsuccessful modification can be a corn-cell clusters in a petri dish that has a gene that makes them resistant to herbicide in the dish. The other cluster does not because only the resistant cells will be growing.

Andrew D. Anderson

The Risk of Moving Genes

Genetic engineers can transfer just a few genes at a time between species that are distantly related or not related at all.

What is it about technological development that makes it such a mixed blessing today that leads to such widespread wariness on the part of the public? How can we understand this double-edge quality of technological development in ways that will help us avoid some of the pitfall of the past? What can we do about engineering education, engineering practice, and public policy to help resolve the paradox and reduce the chances of creating new problems in the future? Certain characteristics of technological development have very definitely caused us problems in the past, both in terms of practice and in terms of perception.

More recent technological developments are incremental in their intended beneficial consequences. This was less frequently the case in the earlier stages of development when the benefits of a new technology, such as electrical energy distribution systems, were dramatic in their effect. For most of human history, the impacts of development were masked and diluted because that development were orders of magnitude away from stretching the capacity of our environment to absorb pollution and other burdens. We did not know what smog was, there were large numbers of rain forest, healthy fowl, wild animals, and we had good fresh air.

We had roots that went a long way back. The slash and burn agriculture required new land every few years. It was never seen as an obstacle or consideration, because land was available without limit, and people were so few. Air pollution in industrial England had severe local effects in the Killing Smog's of London, but those problems seemed not to have significant global consequences and were largely dealt with locally.

How Safe Is Our Food?

People got sick off of contaminated parsley and scallions, cantaloupes, leaf lettuce, sprouts, orange juice, and almonds. Refrigerated potato salad, eggs, chicken, salami, beans, hot dogs, hamburgers, and deli meats caused sickness. It was served in restaurants, nursing homes, on cruise ships, farms, at churches and temples, family reunions, country fairs, casinos, day-care centers and was distributed among many towns, in many states.

The Center for Disease Control and Prevention (CDCP) says each year in the United States 76 million people suffer from food borne disease, 325,000 of them are hospitalized and 5,000 die. In the developing world, contaminated food and water kill almost two million children a year. It usually affects the very young and the very old who have suffered debilitation, disease from what most of us consider one of life's less risky activities, eating.

When you look at this, you naturally think, "risk" should not be in the same sentence with food. In recent years we have heard about the dangerous adulterants contaminating our food, pesticides on our grapes, carcinogens on our strawberries, chemicals on our apples, and poisonous metal in our fish.

Among the agencies that oversee the safety of the United States food supply is the Department of Agriculture charged with regulating meat, poultry, and foods that contain them. It also regulates pasteurized egg products. The Food and Drug Administration addresses the safety of all other foods, including fresh produce, canned and imported foods, milk, shell eggs, seafood, and any processed foods that do not contain meat and poultry.

Those agencies post periodic alerts about hazards in food, chemical contaminants, food additives, and unlabeled allergenic ingredients. But most government officials and health experts agree, the greatest hazards today in the American food supply are not pesticide residues or dioxins or even hidden allergens which are very important, but food borne pathogens, bacteria, virus, and parasites with the potential to harm or kill us.

Most of us get weak from a feverish sweat, suffer abdominal cramps, and diarrhea. The short term ailments of our alimentary tracts are typically caused by viruses, often food borne, and can spread from one person to another by what is called or known as the fecal-oral route (contact with human waste and unwashed hands).

There are 200 times as many bacteria in the colon of a single human as there are human beings who have ever lived. Most of these microbes coexist peaceful with our own cells and even assist them, helping with digestion, synthesizing vitamins, shaping the immune system, and fostering general health. Nearly all raw food harbors bacteria, but the microbes that produce food borne illness are bugs of a different order, capable of causing severe illness and even lasting damage, disorders ranging from temporary paralysis to kidney disease.

Many of these microbes are present in the animals we raise for food. When a food animal containing pathogens is slaughtered, its stomach contents or manure can taint meat during processing. Fruits and vegetables can pick up the pathogens if washed or irrigated with water contaminated with feces of human sewage. A single bacterium in the right conditions, divides rapidly enough to produce colonies of billions over the course of a day, and even only lightly contaminated food can become highly infectious. They can also hide and multiply on sponges, dish towels, cutting boards, sinks, knives, and countertops, where they can be easily transferred to food or hands.

There are new troubling clusters caused by bacteria with unwisely names. Some are new forms and others are the same as always but are in new places. The food contaminated with this nasty set of pathogens tend to look, smell, and taste normal, and the offending microbes, we are learning, can survive the traditional heating and cooling techniques we once thought did away with them.

I remember when growing up in Kansas I would eat cookie or cake dough which was a sweet melting mix of butter, sugar and raw eggs and I licked the bowl. It was thought that if I could avoid only raw eggs with cracked shells which might allow pathogens in I was safe. Now food experts agree that even a perfect egg may not be safe and those with weaken immune systems, life threatening infections, can get inside the ovaries of a laying hen and contaminate her eggs before the shells are formed. This will affect many cookbooks that have been printed.

Organisms in deli meats, smoked fish, blue cheese or soft cheese multiplies at refrigerator temperatures. In one study the microbe turned up on the inside surface of the refrigerators of two-thirds of the patients infected with listeria, the disease. It does not always get into our food, but when it does, it can cause encephalitis or meningitis in people with vulnerable immune systems and, in pregnant women, miscarriage or still birth.

In Philadelphia in October 2002, Wampler Foods issued a national recall of all cooked deli products made since May at an Auburban plant and halted production because the meat was possibly contaminated with listeria. The recall is the largest in United States Department of Agriculture history totaling about 27.4 million pounds of meat. Two hundred ninety five thousand pounds of chicken and turkey products produced at the Franconia plant was also recalled in October of this year.

The company recalled all cooked deli products made from May 1 through October 11 and halted production at the facility which is about 25 miles north of Philadelphia after they received test results of samples taken from floor drains.

A scientific investigation into the cause of illnesses, miscarriages, and deaths in the Northeast from the listeria strain was the result a federal agency said. David Van Hoose, Whampler's chief executive officer said none of their products have been linked to the outbreak. At least 20 death and 120 illnesses have been caused by listeria in eight Northeast states since last summer 2000.

Hamburger can cause nausea, stiffness, high fever, severe headache, and can be fatal in young children, the elderly people with weak immune systems, and cause miscarriages and stillbirths.

Tom Brayton in Orange City, Iowa says it is too risky eating a hamburger after his 20-month old son died from an E. coli infection after a recall of millions of pounds of ground beef from stores nationwide.

There are 1,500 beef slaughtering and grinding plants that submit plans to the agency every year, but now the government says they will come under closer scrutiny than in the past. The General Accounting Office found that only about 1 percent of all Hazard Analysis and Critical Control Point plans undergo stringent scientific review.

If packers find cow's feces before grinding meats, they try to remove it and clean the beef by using hot steam or washing it with acids. But grinding meat from different steers or different farms can spread the pathogen. A new policy urges packers not to mix beef from separate providers when grinding. How can this method be continually checked?

A recall started the last of September after 57 people got sick and the USDA investigated to see if about 10 other illnesses in four states were linked. The recall was from Cargill's Emmpak Foods Inc. that recalled about 3.4 million pounds of ground beef that was sold in hotels, grocery stores, and restaurants. The company shut down its Milwaukee plant idling 160 workers and it figures the company lost $400,000 each day the plant was closed.

The other bad side is beef processors have spent about $400 million on techniques to stop the problem and Tyson Foods subsidiary IBP Fresh meats Inc. spent more than $100 million on food safety measures. Small companies are forced to sell to big corporations, and giants like Hudson Foods are out of beef altogether after recalling 25 million pounds of beef or 12,500 tones.

It is impossible to block E. coli all the time unless irradiation is used and many consumers do not want that. Irradiation is to treat by exposure to radiation. Grinding tons of beef daily will always entertain E. coli so the best actions is to figure the disease is in the meat, and then use a method or methods to remove it from all meats that will be ground.

A laboratory test of 22 types of lettuce bought at northern California supermarkets found four were contaminated with a toxic rocket-fuel ingredient that has polluted the Colorado River. The water is used to irrigate most of the nation's winter vegetables. Even though the Texas Tech University that paid for the test said the sampling was too small, the results were alarming enough to warrant a broad examination by the Food and Drug Administration. Bill Walker said that is nearly 1 of 5 samples of a common produce item contaminated with a chemical component of rocket fuel makes it significant.

The four lettuce samples contained substantial quantities of perchlorate. A prepackaged variety of organic mixed baby greens had a level of perchorate contamination at least 20 times as high as the amount California considers safe for drinking water. The other three were packaged butter lettuce and radicchio, romaine lettuce and radicchio, and a head of iceberg lettuce. They all had at least five times as much perchlorate as California considers safe for water.

It is a salt used to help power missiles and the space shuttle and may cause health problems. It is known to affect thyroid hormone production which is critical to early brain development. It may be especially dangerous for pregnant women and young children.

Tracking Where The Disease Began

Tracking the nature and mechanism of the contamination of one fruit to find out what went wrong in the foods journey from field to the table. In January of 2000, public health officials of Virginia noted an unusual cluster of patients sick with food poisoning from one strain of Salmonella. Fifteen had been hospitalized with severe bloody diarrhea, two had died and the common factor was all had eaten mangoes during the previous November and December.

The investigation of the implicated fruit led to a single large mango farm in Brazil. When a team of health officials visited the farm, they discovered that tanks used to dip the mangoes in warm water to control fruit fly infestation, and then in cold water to cool the fruit, were open to the air. There were toads and birds around the tanks and feces in the water. The cold rinse caused the mangoes to absorb the tank water and the pathogens it contained, including a strain of Salmonella.

This could happen at any time since we demand strawberries, peaches, mangoes, and lettuce the year round and we are depending more and more on imports. Over 40 percent of all fresh fruit consumed in the United States comes from Mexico, Chile, Guatemala, Costa Rica, and other foreign countries, traveling hundreds or even thousands of miles to reach our grocery-store shelves.

Eating food grown elsewhere in the world means depending on their soil, water, and sanitation conditions in those places and on the way their workers farm, harvest, process, and transport the products. It also happens in America. Almonds from a farm in California infected 160 Canadians with Salmonella. Because of the globalization of our food supply, the health hazards of one nation easily become those of another.

Unknown Problems

We like our foods prepackaged and ready to-eat making it much easier for the housewife to fix a meal, but we do not realize the cost we pay. We are leaving to commercial food makers the peeling, chopping, and mixing of our food. We are buying lettuce in plastic packages, potato salad, and humus in deli containers. We are eating out more meaning forty cents out of every United States dollar spent on food is spent outside the home in restaurants and other commercial food services where young or inexperienced, and often underpaid, workers are preparing our meals raising our risks of food poisoning. With more untrained people handling food, the greater the risk of inadequate cooking or of cross-contamination of safe foods from unsafe or uncooked foods.

We like our food cheap and the American farmer wants to make more money. The small plants are genetically raised for a bigger plant so the farmer can make more money and some of the largest and most serious outbreaks of food borne illness have resulted not from imported foods but from the factories and farms within our own borders which provide food to large numbers of consumers.

A case involving contaminated ice cream constitutes one of the largest outbreaks ever recorded in 1994. The premix for Schwan's was being transported by trucks, which is a widely distributed brand of ice cream, carried traces of raw eggs contaminated with Salmonella enteritis. The outbreak sickened an estimated 224,000 people in 48 states.

Another deadly outbreak on record involved various brands of hot dogs and cold cuts made with meat from a Sara Lee Processor. The microbe, an unusual strain of listeria, sickened scores of consumers in 1998 and was linked to 15 deaths and 6 miscarriages or stillbirths. It ended after the company recalled 15 million pounds of meat, one of the largest meat recalls in United States history.

Will Biotech Foods Harm The Environment?

The main safety issues of genetically engineered crops involve not people, but the environment. We are following the same pattern we used in nuclear power and nuclear waste. We worked with it, developed it, storage it, released it in the air, and made a bomb from it without any thought on how to dispose of it. We talk about one half life, active life, and have accumulated many thousands of tons with no place to put it.

Now we are during similar things on biotech foods not having any idea what will be affected and for how long. Allison Snow is an Ecologist at Ohio State University and during research on gene flow which is the movement of genes via pollen and seeds from one population of plants to another. She is concern that genetically engineered crops are being developed too quickly and released on millions of areas of farm land before they have been adequately tested for their long term ecological impact.

What might be the effect of these engineered plants on so called non-target organisms and creatures that visit them? Crops with built-in insecticides might damage wild life were inflamed in 1999 by the report of a study suggesting that Bt corn pollen harmed monarch butterfly caterpillars. They do not feed on the corn pollen, but they feed on the leaves of milkweed plants which often grow in and around cornfields. Test showed that Bt corn pollen dusted onto milkweed leaves stunted or killed some of the monarch caterpillars that were on the leaves.

Follow up studies in the field reported that pollen densities from Bt corn rarely reach damaging levels on milkweed, even when monarchs were feeding on plants within a cornfield. Will wind blow the pollen or how many years will pass before the pollen is on the milkweed? Perhaps a bigger concern has to do with insect evolution. Crops that continuously make Bt may hasten the evolution of insects impervious to the pesticide. Such a breed of insect, by becoming resistant to Bt, would rob many farmers of one of their safest, most environmental friendly tools for fighting the pests.

The United States government regulators working with biotech companies have devised special measures for farmers who grow Bt crops. Farmers must have a moat or refuge of conventional crops near their engineered crops. The idea is to prevent two resistant bugs from mating. The few insects that emerge from Bt fields resistant to the insecticide would mate with their nonresistant neighbors living on conventional crops nearby. The results could be offspring susceptible to Bt. The theory is that if growers follow requirements, it will take "longer" for insects to develop resistance, but they will develop the resistance in time.

Many ecologists believe that the most damaging environmental impact of biotech crops may be gene flow. Could transgressors that confer resistance to insects, disease, or harsh growing conditions give weeds a competitive advantage, allowing them to grow rampantly?

Gene flow from crops to weeds occurs all the time when pollen is transported by wind, bees, and other pollinators, says Allison Snow. There is no doubt that transgenes will jump from engineered crops onto nearby relatives. But since gene flow usually takes place only between closely related species, and since most major United States crops do not have close relatives growing nearby, it is extremely unlikely that gene flow will occur to create problem weeds. Snow says even a very low probable event could occur when you are talking about thousands of acres planted with food crops. And in developing countries, where staple crops are more frequently planted near wild relatives, the risk of transgenes escaping is higher. However, she thinks it will not be long before super-weeds will emerge.

Many ecologists believe that industry should step up the extent and rigor of its testing and governments should strengthen their regulatory regimes to more fully address environmental effects. Every transgenic organism brings with it a different set of potential risks and benefits. Each needs to be evaluated on a case-by-case basis. But right now only one percent of USDA biotech research money goes to risk assessment.

How Biotechnology Can Help

Plant biotechnology will likely become even more important in creating a greener 21st century. An increasing global population is

225

fueling a demand for more and better food. Experts say farmers will
need to at least double their production over the next 25 years to feed
these new mouths, at a time when annual yield increases have slowed.

That put both agricultural and wilderness areas under intense
pressure. Environmental experts fear that up to half the world's 6
billion acres of tropical forests will be lost to agricultural expansion.
Annual plowing causes "two billion tons" of United States topsoil to
erode into rivers and eventually into the Gulf of Mexico where soil
and fertilizers leave a dead zone. Biologists warn that as much as 20
percent of all species in those forests could be extinct within 30 years
and two out of every three people could live in water-stressed
conditions by 2025.

One Genetic group cut a stalk of green corn level to the
ground they had experimented with and placed it in a container of
water. It remained green for seven months. The group is trying to get
the roots to remain alive in the soil and grow the next season. Thus it
will help prevent the topsoil from eroding into rivers and the Gulf of
Mexico.

Biotechnology could help by helping farmers grow more food
on existing acres, including drought-prone or other marginal lands,
which could reduce the need to put remaining wilderness areas under
the plow. Researchers are developing corn and rice plants that are
more tolerant of aluminum, a common soil toxin and tomatoes and
other crops that can thrive in salty soil that is an agricultural problem
in many regions where irrigation is used and it will continue to get
worse.

Biotechnology is helping farmers produce more corn that can
be used for bio-based fuels such as ethanol.   In the future,
biotechnology could help develop more renewable raw materials for
energy and other industrial uses and provide even more environmental
benefits.

Eight biotech crops planted in 2001 including soybeans and
cotton, boosted total production by 4 billion pounds, reduced spraying

by 46 million pounds, and generated an additional $1.5 billion in income for farmers.

We must realize that only six to eight inches of topsoil is all that stands between much of the world and starvation. Yet each year large plots of agricultural land is lost to salinization, erosion and other forms of soil degradation. Less topsoil means less food and degraded soils have lower global yields. That is one of the problems in all developing nations.

Since we are not ready to accept starvation, or place parks and the Amazon Basin under the plow, there appears to be only one good alternative, discover ways to increase food production from existing resources. Biotechnology maybe the only way to increase production to future food needs said Martina Newell-McGloughlin director of the University of California System Wide Biotechnology Research and Education Program.

Organic Food

New government rules will define organic. The sale of these fruits, veggies and snack foods has soared, but we still are not sure what good they do. Here is a guide to how purer products affect the health of our families and the planet without the employment of chemically formulated fertilizers or pesticides.

Organic farms provide less than 2 percent of the nation's food supply and take up less than 1 percent of its cropland, but it is flourishing. The market for organic food has grown by 15 to 20 percent every year, five times faster than food sales in general. Nearly 40 percent of United States consumers now reach occasionally for something labeled organic and sales are expected to top $11 billion this year. The new organic is all about bigger farms, heartier crops, better distribution and slicker packaging and promotion.

Big companies like Heinz and General Mills are now launching or buying organic lines, and selling them in main stream supermarkets. As of October 21, 2002 any food sold as organic had to meet criteria set by the US Department of Agriculture. The foods must be produced without hormones, antibiotics, herbicides, insecticides, chemical fertilizers, genetic modification or germ-killing radiation. From now on, consumers will get a very solid idea of what is organic and what is not.

Peter Hoffman is owner of New York's Restaurant Savoy and chairman of the Chefs' Collaborative. When people taste asparagus or string beans grown in richly composted soil, they cannot get over the depth and vibrancy of the flavor. To most consumers, organic means healthier.

Eggs and milk were among the fastest, growing organic categories during the 1990s. Dairy sales rose by 500 percent from 1994 to 1999. Eggs from some free range chickens may contain more beneficial omega, 3 fatty acids. One hundred percent organic products carrying this label cannot contain any non-organic ingredients. Ninety five percent of the product's ingredients must be organic.

Organic growers, with their smaller harvests and their reliance on nearby markets, can plant delicate heirloom strains and give the fruit more time on the vine without spraying it. They pick it when it is ripe. Under new guidelines, meat and dairy products must be labeled organic and must come from creatures that are raised on organic grains or grasses that give access to the outdoors. Organic food is tastier and more appealing.

As demand increased more farmers are discovering they can enrich the soil and manage some pest simply by rotating their crops. They are learning that they can often control insects without other insects, or lure them away from cash crops by planting things they prefer. Well managed organic farms often match conventional ones for productivity and beat them when water is scarce. The organic method may be the key to part of our survival.

Chapter 14: Are We Preparing for The Future?

A Look at Ourselves and Our Actions

In North America not only does the Colorado River barely make it to the Gulf of California, but last year even the Rio Grande dried up before it merged with the Gulf of Mexico. In Central Asia the Arial Sea shrunk by half after the Soviet's began diverting water for cotton and other crops. China's river history of ruinous floods, the Yellow River now barely trickles in its lower reaches-and in recent years has gone dry due largely to heavy irrigation upstream. It is not alone, the once mighty Nile, Ganges, barely reach the sea in dry seasons.

As mentioned earlier, Lake Chad in Africa in 1962 was the fourth largest water body on the continent, a shallow 9,600 square-mile expanse tucked between the Sahara and the rain forests of central Africa. Now it has shrunk to 20th of that size. Drought since the late 1960s led to overgrazing on surrounding grassland. A large scale irrigation project designed to distribute water from the lake failed when the lake's level dropped below the intake location.

The Colorado River is the lifeblood of the booming, if parched, American Southwest. Seven states and Mexico use nearly every drop, reducing one lush delta on the Gulf of California to a sliver in a sun baked mudflat. The river fills the thirst of more than 25 million people and it irrigates some of the nation's most profitable farms in California's Imperial Valley, but California has long taken more than its share. It will soon have to decide whether to water its farms or its burgeoning cities.

Pumping To Much Water

Elsewhere, countless small rivers have gone dry. That agricultural achievement, which has enabled the country to grow enough food for its one billion people, was accomplished because of a huge increase in groundwater pumping. In the mid-fifties fewer than 100,000 motorized pumps were extracting groundwater for agriculture. Today about 20 million are in operation, with the number growing by half a million each year.

With farmers extracting water more quickly than nature replenish it, aquifers have been depleted to the point that roughly half of India now faces over pumping problems, such as ground water shortages or the influx of salt water into coastal wells. Many farmers have been forced to abandon wells or keep drilling deeper, raising costs and driving some out of business. In parts of Gujarate the water has been dropping as much as 20 feet a year.

One reason farmers in India and throughout the world have been needlessly pumping water is that they pay so little for it. In India the water itself is free, and the government heavily subsidizes the electricity that drives the pumps. Rather than pay for the number of hours a pump runs, farmers pay a low flat annual rate and pump whenever they wish. The over pumping of, aquifers, whether for agricultural or municipal use, extends far beyond India.

The water under the North China Plain, which produces about half of China's wheat and corn, is steadily dropping. Sandra Pastel, a freshwater expert and director of the Massachusetts-Based Global Water Policy Project said, that continuing groundwater depletion could reduce China's and India's grain production by 10 to 20 percent in the coming decades.

American farmers are withdrawing water from the Ogallala aquifer, which underlies the Great Plans, at an unsustainable rate, with a third of the Texas portion already significantly depleted.

The 20th century witnessed the greatest expansion of water control in history, thousands of miles of canals, but at the end of all that shifting of earth and water we find ourselves contemplating a global water crisis. Thanks to our growing population and the growth in agriculture and industry that has accompanied it, our water sources have been rendered increasingly dirty and depleted. Half the world's rivers are either polluted or running dry.

Unchecked irrigation poses a serious threat to rivers, wetlands, and lakes. China's Yellow River, siphoned off by farmers and cities, have failed to reach the sea most years during the past decade. In North America not only does the Colorado River barely make it to the Gulf of California, but last year even the Rio Grande dried up before it merged with the Gulf of Mexico. In Central Asia the Aral Sea shrunk by half after the Soviets began diverting water for cotton and other crops. Elsewhere, countless small rivers have gone dry.

Israel may be heading toward a war over water with Lebanon. The United States is sending an envoy to the parched region to help resolve an increasingly dangerous dispute over a river called the Wazzani. Paul Patin said we are discussing the issue at high levels in Israel, Beirut, and Washington that is a spokesman for the US Embassy in Tel Aviv.

There is a new plan by Lebanon to tap a river that is a critical source of drinking water for Israel. The springs flow into the Hasbani River, which flows into the Jordan River and on to Israel's Sea of Galilee. The sea, actually a big lake, is Israel's most important source of fresh water.

Water rights represent one of the most explosive issues in an arid region where water for drinking, agriculture, and sanitation is in desperately short supply. Most of the wars fought between Israel and its Arab neighbors since the formation of the Jewish state in 1948 have been at least partly provoked by water disputes.

In summer the taps run only a day or two a week in Amman, Jordan, so residents have to store water in rooftop tanks. The rations will likely get even tighter in years to come, and Jordan's population is on track to double within a quarter of a century.

When Jordan announced plans to divert 1.3 million gallons per day from the Wassani to provide drinking water for several villages,

231

Israel's reaction was swift. Prime Minister Ariel Saron declared that Israel would regard such an action by Lebanon as "a pretext for war." A high ranking official said that if Lebanon initiates an engineering project to divert the water, "Israel would be forced to take action and I mean direct action to protect our precious water."

The Lebanese government insisted that the diversion plan is in keeping with international law, saying Israel's only claim to the water is based on its 22-year-long military occupation of southern Lebanon that ended in 2000.

Drought stalks parts of North America, Europe, Africa, North and Central Asia, and Central America as a warming climate makes rainfall patterns less predictable. There have been water riots in Colombia, Bolivia, India, Bangladesh, and Pakistan. As many as 6,000 children die every day from water-related diseases.

And our water troubles are just beginning. Christine Todd Whitman admits to being puzzled by the state of our waters. In a March meeting, she called water our single greatest challenge, "I do not know how we are going to solve it all."

As we confront the bold fact that the world's physical supply of water has limits, we dig our wells deeper, remove salt from ocean water at huge expense, and use and reuse water over and over. We compete for control of shrinking rivers and move large amounts of water longer and longer distances in strange conveyances. Italy has undertaken feasibility studies for an underwater pipeline to bring water from Albania. Las Vegas officials shop for water across the Rockies in Wyoming, and representative of sheiks in Abu Dhabi have offered to build dams in countries away in the mountains of Pakistana, in order to shuttle mountain water south to their desert kingdom.

The US Jet Propulsion Laboratory is developing a zeppelin that can carry as much as 50,000 tons of water. Its creators believe it can be used to put out forest fires or airlift a small iceberg to drought-stricken regions. Canada's Global Water Corp signed an agreement to ship 5 billion balloons a year from Stika's Blue Lake all the way to China, until the Canadian government, wary of losing control of its bountiful resource, slapped a ban on bulk exports of water.

Ideas that seemed ridiculous only a few years ago, such as towing icebergs south, now seem plausible. In Chile, people are harvesting clouds. University of Chile researchers are able to catch coastal fogs in vast nets of polypropylene that trap moister and collect enough clean fresh water to supply whole villages. Fact and science fiction draw closer to one another in schemes like Israel's plans to build a canal from the Mediterranean to the Dead Sea, using the fall in sea level to desalinate water and generate hydropower, or its agreement with Turkey to buy large amounts of water to be towed across the Mediterranean in large plastic barges. Our credibility is strained by a proposal to create a giant lake in central Africa to bring rainfall to the Sahel, to build a dam across the Bering Strait to control cold-water currents that will then warm the air of Alaska and Siberia, or to turn the Grand Canyon into a storage reservoir.

For most of us, the answers to our water problems are far more practical. They are to be found in the reuse of urban and industrial water, of conservation and careful use in agriculture, fixing leaks, clean up technologies, desalination plants, smart regulation, and restoration of watersheds and aquifers. Turkey's dams on the Tigris and Euphrates River that have already begun to affect Syria and Iraq, threatens disaster. How well we manage our water all over the world will write the story of the 21st century. We need to plan for our aquifers, lakes, and rivers before a crisis overtakes us, to figure how to conserve them, keep them clean, and let our political leaders know that we expect them to consider the future impact of their decisions.

We need to make it our business to know where our water comes from, to understand that the weed killer we put on our lawns, the chemicals used on our crops, or to keep ice off the intestates, will all eventually make their way into our groundwater and rivers. We need to realize that the oceans are immense and beautiful, but we do make an impact on them, The United States need to work to lose its appetite for oil, stop building mega-dams that alter the flow of rivers and drown land, stop over fishing and discover ways to help our water systems.

Is Water Out Of The Future?

The largest man-made edifice built in China since the Great Wall is the Three Gorges dam. Beijing's leaders believe the dam will cure many of the nation's ills generating jobs, commerce, and a ninth of China's electricity. To its legion of critics, the 1.3 mile long 600-foot-high monolith is a 25-billion-dollar environmental nightmare. It will displace nearly two million people, inundate some of China's best farmland, and create a 370-mile-long lake that critics say will fill with a billion tons of sewage and industrial wastes each year. If the river's silt builds behind the dam, its capacity for hydropower and flood control will be diminished.

The Murraya-Darling basin waters Australia's bread basket covers one-seventh of the country, producing 41 percent of agricultural income and supplying drinking water for two million people. But two centuries of clearing native vegetation for agriculture and extensive irrigation have raised water tables in the region, bringing the soil's naturally occurring salts into roots, zones, and rivers. More than 1.6 million acres are seriously affected by salt. That could spread to nearly 8.4 million acres by 2050, despite salt reduction projects built to protect the rivers. Basin salinization costs Australia more than 170 million dollars a year in water treatment and lost productivity, leading officials to call it the most important environmental issue facing the nation.

The Mesopotamians first used water from the Tigris and Euphrates Rivers more than 7,000 years ago. Today Turkey, Syria, and Iraq have built more than 30 dams on the rivers, enough to capture their flow many times over. Iraq has threatened to bomb upstream dams. Syria and Turkey have come close to blows over a Turkish irrigation project. Wyria and Iraq claim water use rights dating from prehistory, and Turkey demands sovereignty over the headwaters in its land.

On the outskirts of Delhi, India, squatters live in the shadow of a big water pipe carrying water to the city. Where there is a leak, squatters use it to take a shower. Delhi's population is growing so fast its water supply cannot keep up. Only half the city's inhabitants can get treated water in their homes.

Population is booming as water pipes crumble, and Mexico City must truck water to many residents. Once called the Venice of the New World, the city has so drained its aquifer since 1900 that its water has sunk two dozen feet. As the ground shifts, pipes break causing leaks that claim nearly a-third of its water.

In Las Vegas, the driest state from rain fall with excess and illusions, water is no exception. At the Bellangio Hotel, 27 million gallons of water dance to show tunes through choreographed nozzles in an artificial lake. With nearby Hoover Damp providing precious Colorado River water, the Las Vegas residents all have pretty green lawns.

All over the globe farmers and municipalities are pumping water out of the ground faster than it can be replenished. The United Nations said that 2.7 billion people would face severe water shortages by 2025 if consumption continues at the current rates. Fears about the future arise from a projected growth of world population from more than six billion today to an estimated nine billion in 2050. Yet the amount of fresh water on earth is not increasing. Nearly 97 percent of the planet's water is salt water in seas and oceans. Close to 2 percent of earth's water is frozen in polar ice sheets and glaciers, and only a fraction of one percent is available for drinking, irrigation, and industrial use.

Nearly 70 percent of the world's fresh water is frozen in glaciers, permanent snow cover, ice, and permafrost. The Antarctic and Greenland ice sheets have the bulk of it. Groundwater, thirty percent of all fresh water is underground, most of it in deep, hard to reach aquifers. Together lakes and rivers contain just a little more than one-fourth of one percent of all fresh water. Lakes have most of it with soils and wetlands. Though clouds and water vapor hold just four one-hundredths of one percent of all fresh water, they still have six times more water than all the world's rivers.

United States Rated World's Highest Wasteful Water User

Some of the world's richest countries, including the United States and Japan, lag behind some developing nations in making the best use of water, according to a new grading system published in 2002. The United States was rated the world's most wasteful user of water by the first Water Poverty Index. Finland was ranked highest on the index, which graded 147 countries according to resources, access, capacity, use, and environmental impact. The rest of the top 10 were Canada, Iceland, Austria, Norway, Sweden, Guyana, Suriname, Switzerland, and Ireland. The ten countries at the bottom of the index were Burundi, Rwanda, Benin, Chad, Djibouti, Malawi, Eritrea, Ethiopia, Niger, and Haiti. Issues raised by the index are to be discussed in March, 2003 at the World Water Forum in Japan.

The links between poverty, social deprivation, environmental integrity, water availability, and health become clearer in the index, enabling policy makers and stakeholders to identify where problems exist and the appropriate measures to deal with their causes said Caroline Sullivan, who led the team developing the Water Poverty Index at the Center for Ecology and Hydrology in Wallingford, England. The center is part of the British government funded natural Environment Research Council.

One fifth of the world's population in 30 countries faced water shortages in 2000, a figure that will increase to 30 percent of the population in 50 countries by 2025, according to the World Water Council based in Marseilles, France.

Water demand is increasing three times as fast as the population growth rate even through no new water can be created anywhere on this planet, said World Water Council president Mahmoud Abu Zeidd.

Ireland, Iceland, Japan, Spain, and Austria were rated tops in the capacity category, which defines a country's ability to purchase, manage, and lobby for improved water, education, and health. The bottoms five were Sierra Leone, Niger, Guinea-Bissau, Mall, and the Central African Republic, some of the world's poorest nations. The United States was ranked 32nd overall in the index, but last in efficiency. William Cosgrove of the world Water council said the United States is at a relatively low position because of wasteful or inefficient water use practices in domestic, industry, and agriculture. This is illustrated that their per capita water consumption is the "highest in the world." Japan ranked 34th with a low score on environmental factors.

Beyond The Horizon

Imagine a future of relentless storms and floods, islands and heavily inhabited coastal regions inundated by rising sea levels, fertile soils rendered barren by drought and the desert's advance, mass migrations of environmental refugees, and armed conflicts over water and other precious natural resources.

One might just as easily conjure a more hopeful picture of green technologies, livable cities, energy efficiency homes, transport and industry, and rising standards of living for the world's people, not just a fortunate minority. The challenge of living in harmony with the earth is as old as human society itself. That relationship changed fundamentally, a little more than one century ago with the industrial revolution.

Today we need another revolution, a revolution in our sense of global stewardship. For too long, too many people have believed that natural limits to human well-being have been conquered. And too many have put their faith in technological break through as the inevitable answer to any resource constraints or other vulnerabilities that might emerge.

Slowly as human kind find itself in uncharted territory with respect to energy use and population growth, and in particular the natural desire of all people to share the prosperity so far enjoyed by only a few, we will began to recognize the perils inherent in the prevailing model of development. As forest have been cut and aquifers over pumped, as the atmosphere has filled with toxins, the oceans have been fished to exhaustion, and as the climate itself has begun to react, we should grow wiser.

Chapter 15: Beginning of Our Nuclear Age

Writing about our pollution, our threat to good health, and our cancer causing products, we will end this book on some of the unsolvable problems with nuclear waste.

This year in 2003 the world's commercial nuclear reactors will create more than 11,000 tons of radioactive spent fuel. The waste poses a risk in the form of leakage, accidental leakage, and as a terrorist target. The United States is home for nearly a quarter of the world's reactors and 161 million people live within 75 miles of an above ground waste storage site. Pending national debate and more studies, 131 sites in 39 states may send a portion of their waste to be entombed beneath Yucca Mountain in Nevada starting in 2010, but there will be millions of tones that will not be safely stored and will never be stored.

Nuclear Waste Is Radioactive

The nuclear age dawned almost fifty years ago when the Atomic Bomb exploded into our consciousness. Today industrial nations use nuclear energy to produce weapons and supply some 15 percent of the world's electricity. Compared to fossil fuels, nuclear energy is cheap, clean, and efficient, but it also comes with a deadly by-product, nuclear waste. We have not learned how and where to store it safely for a long time or a short time period.

Radioactivity is the property possessed by some elements, as uranium, of spontaneously emitting alpha, beta, or gamma rays by the disintegration of the nuclei of atoms. Alpha rays or an alpha particle is radioactive moving at a very high speed. All three are a stream of particles moving at different very high speeds during radioactive decay. It can sicken, kill, and cannot be burned. It cannot be buried underground safety, released into the atmosphere safely, and it cannot be ignored, because every year the amount of nuclear waste generated by over 100 American nuclear reactors increase.

It consists of by-products and extra materials that are generated and left over during the various processes that go into the making of nuclear energy. Some waste materials are produced in the mining of nuclear ores, and some are generated in reactors during the fission process. Others take shape during the development and manufacture of nuclear weapons, and in the research aimed at improving nuclear equipment and production techniques. Others are from the preparation of nuclear medicines. They all present the same problem, contamination by radioactivity.

How does material become contaminated? Rods used in the fission process have been saturated with high-powered radioactivity. These rods are stainless steel or zirconium cylinders in which uranium pellets are placed for fission. The rods measure 3 to 14 feet long and about a half inch in diameter. They are bundled together in units of 30 to 300 rods and inserted into the reactor core. Each bundle is known as a fuel assembly.

Various numbers of fuel assemblies are placed in reactors, depending on the size. In time, the assemblies become so shot through with radioactivity that they can no longer perform efficiently and must be replaced, usually every two to three years. The replaced assemblies, spent fuel rods, and their uranium contents are listed as wastes and are called spent fuel assemblies, spent fuel rods, and spent fuels. The assemblies and rods are highly dangerous wastes because they are saturated with the various radio-nuclides that recreated during fission.

Example of waste are, protective clothing worn by nuclear workers, gloves, shoes, masks, hand tools of all kind, wiping rags, and such personal possessions as handkerchiefs. Laboratory test tubes, and the carcasses of animals used in nuclear research are also dangerous.

Sources of Radiation

The largest doses of radiation from non-natural sources come from radiation used for medical diagnosis and treatment, color televisions and video games can account for a considerable amount of man made radiation.

Nuclear industry, consumer products, nuclear medicine, rocks and soil, medical X-rays, and natural background radon also contribute. The largest doses of radiation from non-natural sources come from radiation used for medical diagnosis and treatment.

Nuclear energy processes produce three basic type of radioactive waste.

1. Uranium mill tailings are the earthen residues, usually in the form of fine sand after the mining and extraction of uranium from ores. These wastes emit low levels of radiation, mostly in the form of radon, which can contaminate water and air.

2. Low-level waste contains varying lesser levels of radioactivity, including trash, clothing, and hardware. Spent fuel is used reactor fuel that will be classified as waste if not reprocessed and is usual uranium and plutonium. High-level waste is the by-product of a reprocessing plant. These wastes contain highly toxic and extremely dangerous fission products that require great care in disposal.

3. Transuranic wastes are man-made radioactive elements with an atomic number larger than that of uranium 92, and half-lives of thousands of years. They are in trash produced mainly by the nuclear weapons plants and are a part of the problem that must be directly resolved by the government.

Prior to the early 1970s, the tailings were believed to have such low levels of radiation that they were not harmful to humans. Miners, many of whom were Native Americans, received little protection from the radiation. Now, many of them are reporting very high rates of cancer.

Tailings were also left in scattered piles without warnings of safeguards, exposing anyone that came near. Some tailings were deposited in landfills, and homes were built on top of them. They were generated in large volumes, about 10 to 15 million tons annually. For 15 million tons annually that's over 41,200 tons per day.

Radium-226, the major radioactive waste product, retains its radioactivity for thousands of years and produces two potentially hazardous radiation conditions, gamma radiation and the emission of gaseous radon. There is a proven causal relationship between these radioactive elements for leukemia and lung cancer. Congress passed the Uranium Mill Tailing Radiation Control Act of 1978 which regulated mill tailing operations.

Low-level radioactive waste decays in 10 to 100 years. Until the 1960s the United States dumped low-level-waste, LLW, into the ocean. The first commercial site to house such waste was opened in 1962 and by 1971 six sites were licensed for disposal. The low Level Radioactive Waste Policy Act was passed in 1980.

Storage problems for hazardous waste are many because buffer zones of land surrounding each site are required. This acreage will require monitoring and limiting land use applications for at least a century. Degradation of the packages that contain stored waste can occur from temperature fluctuations, corrosion, water, and containers become brittle. With the amounts of waste accumulating, illegal dumping began and will continue.

Three Types of Radiation

Each radioisotope emits a characteristic pattern of one or more of three types of radiation, alpha, beta, and gamma radiation. Alpha radiation can do the most severe localized damage to living tissue, but it is the least penetrating. It cannot pass through human skin, even a piece of paper wilt stop it. But if a lone-live radioisotope that emits alpha radiation is inhaled, swallow, or gets into a wound, even a tiny speck can do terrible damage. The longer it remains embedded in living tissue, the more harm it will cause.

Alpha radiation is emitted by the decay products of some radioactive elements found in nature, uranium, thorium, and radium, and by long-lived transuranics by products of nuclear technology that is heavier than uranium, such as plutonium.

The nuclear fuel cycle generates hundreds of kinds of radioisotopes and unstable atoms that give off radiation while they decay into more stable structures. Some radioisotopes are more intensely radioactive than others, and some remain radioactive longer than others. Each radioisotope has a distinctive half-life, the length of time a sample of it takes to emit half of its radioactive. Half-lives of radioisotopes range from a fraction of a second to millions of years. As a radioisotope decays, it may turn into yet another radioisotope, called a decay product, which will have its own distinctive half-life and intensity of radiation. So how can we guess the length of the new life or half life?

Beta radiation can penetrate skin, but damage people when it is ingested and the body mistakes it for a chemically similar element. Radioactive strontium is mistaken by the body for calcium. Gamma radiation has still greater penetration power and it can even pass through wood. Most nuclear waste emits beta or gamma radiation, often both.

Radiation harms living creatures by damaging individual atoms inside individual cells. A massive dose of radiation can scramble enough cells so badly that the victim's body can no longer function, leading to death within a few painful weeks. The cumulative impact of these radiation affected cells, multiplying over many years can create a variety of harmful health conditions, from thyroid to genetic disorders to cancers.

No one knows for sure how much radiation is a safe amount to receive. People exposed to high does of radiation die, people exposed to lower radiation get sick, and people exposed to still lower doses suffer no apparent ill effects. All of us are exposed to some radiation. A small amount of natural background radiation seeps from the earth and bombards us from space and we receive additional small does from medical treatments.

Low Level and High Level Wastes

Low-Level Waste, LLW, is generated wherever radioactive materials are used. It is produced in such facilities as reactor plants, industrial plants, hospitals, government, universities, and commercial laboratories devoted to nuclear research and development. It is much less dangerous than High Level Waste, HLW, and it contains radioactive elements with low hazard levels and fast decay rates. Their radiation drops to safe levels in anytime from a few seconds to a few years.

Radiation from some LLW is high enough to require that they be encased in a dense material and shielded from workers for a time. It can usually be safely stored by shallow burial. The burial must be made with care in locations that are controlled so that they will not be used indiscriminately by nuclear workers or the public. It is dangerous when a person touches it, swallows it, or inhales its fumes.

Nuclear wastes have been accumulating over the years because we cannot dispose of it as we do many other industrial wastes. We cannot burn them or wash them away in a river. Burning would only poison the air with their deadly radiation, and washing it away would just as lethally pollute our drinking water.

Nuclear waste or radwaste comes from:

1. Manufacture of nuclear weapons.
2. Extra nuclear material left over.
3. Preparation of nuclear medicines.
4. Research and in improving nuclear equipment.
5. Waste material from mining nuclear ores.
6. Reactors during fission process.

They all have one thing in common, contamination by radioactivity. Assemblies in reactors have a short efficient life that is usually three to four years, but length of time and the wide spread problems they cause are tremendous. Everything it comes in contact with is dangerous including shoes, hand tools, clothing worn, gloves, masks, socks, handkerchiefs, bulldozers, train, and train cars. Also included in that group are liquids and gases. Since their radioactive life "may be long," disposing of it in anyway is never "100 percent."

High-level waste, HLW, is the most dangerous and emits high heat, cannot be handled by humans, and it must be shielded and permanently stored. The decay time can take up to 10,000 years or even more. Scientist really do not know.

High-level waste is generated during the production of nuclear energy for both civilian and military purposes. It generates real heat and emits a dangerous radiation. It requires remote handling, which means that there must be no human contact with it, and it must be stored in a dense material if we are to be safely shielded from its radiation. The materials that do the best job of shielding are lead, concrete, steel, glass, or water. These wastes contain not only an array of very concentrated, long-lived radioactive elements, including plutonium, but also substantial amounts of other hazardous materials such as mercury and solvents.

About 11.3 million cubic feet of the waste lies buried at government sites, and some has been deposited in salt caverns in New Mexico. We can only shield it which means "storing all of the wastes" so that they are secured and shielded beyond our reach. With all of the waste we have, that is impossible to do.

Nuclear waste with invisible radiation intense enough to kill in minutes is stored in drums with spent nuclear reactor fuel and confined in steel cylinders behind glass walls at government facilities in Idaho. Thousands of steel drums are storied there until the United States has a permanent repository for such dangerous wastes.

Uranium mill tailings consist of earthen residues left over from the mining of uranium, ore, and the extraction of uranium from the ore. Made up of soil and fine sand, they contain low-level concentrations of such naturally occurring radioactive substances as radium-226 and thorium-230. The mill tailings emit the radioactive gas radon-222 (which is created by the decay of their radium content) and some gamma rays. As they decay, the wastes also give off alpha and beta radiation.

There are more than "200 million tons" of radioactive mill-tailings in the United States. It is stored "unsafely" in sparsely populated areas of western states as Arizona, New Mexico, Utah, and Wyoming.

Nuclear Waste Sites

Rusting drums of waste are stored at Los Alamos National Laboratory in New Mexico and Sealed Hulls of 92 submarines containing nuclear reactors emptied of fuel are stored at Hanford, WA.

Another Cold War Battlefield, the Idaho National Engineering and Environmental Laboratory west of Idaho Falls, began its career as a firing range for battleship guns in World War II. Later the vast expense of sagebrush and shrub became a research center for nuclear reactors and for a time was used as a permanent repository for some nuclear waste.

Trucks carried transuranic wastes over interstate highways to a burial site in New Mexico, which will be filled with an estimated 850,000 Drums by 2035.

Four double-walled underground tanks had radioactive waste from plutonium processing, Hanford's specialty from 1943 to 1989. In all, tanks held 53 million gallons destined for vitrification in glass, (to divert into glass by heat or fusion). With this much and other lethal waste, Hanford houses the nation's largest concentration of high level nuclear wastes.

What to Do About This Waste

David Lyle is a river guide and talked about a very large mound of uranium tailings close by the Colorado River near his home in Moab, Utah. Because ammonia leaches from the tailings into the river to threaten endangered fish, cancer causing radon wafting from the pile has settled as a radon "fog," said he is tired of looking at ten million tons of tailings.

Ines Triay earned a Ph.D. in chemistry from the University of Miami. She manages the Waste Isolation Pilot Plant in New Mexico, a repository that expects to receive 850,000 drums of transuranic waste by 2035.

Nuclear waste needs to be stored far beyond human reach, where it can decay for the hundreds, and thousands of years necessary for it to become harmless, but how do we devise an effective permanent storage site for nuclear waste, and where do we put it? The issue remains lost in a wilderness of science and politics. Only within the last decade has Congress, by passing the Low-Level Radioactive Waste Policy Act and the Nuclear Waste Policy Act, taken steps to address the problem and has also found that we do not know what to do with it. We discovered we could use it to satisfy our needs, but did not consider that some day we would have to dispose of it.

The nuclear age has been with us since the world's first nuclear explosion in 1945 through the nuclear arms race of the Cold War in the 1950s and 1960s to today's use of nuclear power as an energy source. The technology that created the weaponry and the reactors has not yet devised a solution to one of the most severe problems facing governments and scientists today, what to do with nuclear waste?

Over 38,000 cylinders of depleted uranium, many in poor condition, lie outside the Paducah Gaseous Diffusion Plant which is a center for nuclear-fuel processing. Tainted groundwater has forced the government to provide municipal water for near by residents.

In April of 2003, Yankee Atomic Electric Company boosted its request for damages from the US Department of Energy to $191 million for failing to remove 533 spent fuel rods from the decommissioned nuclear plant. They are trying to force the DOE to be accountable for their 1982 promise to remove the nuclear waste at the Yankee Rowe plant.

Kelly Smith, spokesperson for the Yankee Atomic Electric Company said, the DOE was to have taken possession of Yankee Rowe's waste on January 31, 1998. Yankee Rowe is the third oldest nuclear facility in the United States, shut down its reactor in 1991 because of doubts about the safety of the reactor. The 2,000 acre plant cannot be fully decommissioned until all the waste is removed. It is in the Deerfield River Valley amid some of the steepest mountains in the Berkshires. That location was chosen for cooling purposes, but the area was never studied for use as long term nuclear waste storage.

The Maine Yankee and Connecticut Yankee Nuclear facilities are also seeking claims from the DOE as well as other facilities all across the country. Charles Miller, spokesman for the Department of Justice is handling the suit for the DOE. Neil Sheehan of the Nuclear Regulatory Commission said, it will be years, if ever, before the Yucca Mountain repository in Nevada will be ready to accept nuclear waste.

The word, "cleanup" is used as if after waste is cleaned up by our definition, there will be no more problems. What does the phrase mean? Does it mean that the waste and radiation is 100 percent cleaned up or eighty percent cleaned up?

The National Academy of Sciences Panel recently concluded that DOE, Department Of Energy, should drastically curtail its plans to store plutonium contaminated soil in a new Mexico repository known as the Waste Isolation Pilot Project, until doubts about the sites safety could be answered.

The wastes that pose the most serious immediate risk to workers and the public are liquids. These were largely produced at the Hanford site in Washington State, run by Westinghouse for DOE, and the Savannah River Plant in South Carolina, run by DuPont. Together, the two facilities have generated nearly 100 million gallons of liquid HLW.

Problems with handling these extremely radioactive substances stem from the origin of the nuclear weapons program during World War II. Architects of the Manhattan Project devised the idea of storing the waste in tanks as an interim emergency method until long term solutions could be found. Because stainless steel was then in short supply, officials decided to use carbon steel tanks. But the waste is acidic, which meant it had to be neutralized so it would not dissolve the carbon steel. The neutralization process which involved adding lye and water to the wastes turned out to cause major problems.

Adding water increases the volume of waste, making disposal more difficult. Also, adding lye created chemical reactions that allowed the radioactive elements to precipitate as sludge. Some 90 percent of the radioactivity became concentrated at the bottom of the tanks where heat built up and causes the tanks to crack. At least 500,000 gallons of highly radioactive liquids have already escaped from the Hanford tanks and leaked into the ground because of such corrosion cracking. One hundred forty nine tanks at Hanford are so compromised that the wastes cannot be removed without risking further leaks. In some cases, salts created by the neutralization process are plugging the cracks. If the contents were removed, the tanks would break wide open and leak into the ground.

Although the tank option was supposed to be temporary, managers of the weapons program continued to store high level wastes in carbon tanks and the practice will continue. No time table was set for emptying the tanks and devising a secure long term program for managing the waste.

When the world's first nuclear explosion, code named Trinity, lit up the New Mexico desert on July 16, 1945 it did not ignite the atmosphere, but it spread bits of plutonium and other dangerous radioactive particles hundreds of miles around the test site. The explosion also left behind an inevitable by-product of nuclear weapons production, lots of hazardous waste. Was there a reason the scientist did not know or even suspect there would be radiation that would spread from the initiation site effecting unsuspecting civilians?

Nuclear waste from Trinity, and from the production and explosion of the two bombs that later killed and maimed hundreds of thousands in Hiroshima and Nagasaki in Japan, remain dangerous today, and much of it will remain dangerous for thousands of years into the future. Over the past half century, nuclear weapons production, nuclear power, and other uses of nuclear technology have left the world with mountains of nuclear waste. What to do with it all is a daunting problem indeed.

One Flawed Method for Storing Nuclear Wastes

At Savanna River in South, Carolina, managers decided to allow liquids fresh from the reprocessing plants to decay and cool somewhat, for several months. This material was evaporating, which caused it to be moved to evaporators and then back to the original tanks. The "Tank farms," as mentioned before, have had pumps, jets, and miles of pipe, but much of this equipment has leaked, failed to work properly, waste has spilled and contaminated workers as it was moved from tank-to-tank.

DuPont installed concrete shields with partial steel linings around the tanks at Savannah River. Newer units had a full secondary steel liner. Even this double shell and concrete was not an effective barrier to the hot wastes. Within five years of construction, four of six tanks had developed leaks in both the primary and secondary shields. Other methods used were:

1. Burying dry plutonium-contaminated wastes in "cardboard boxes."
2. Releasing large amounts of radioactive materials in the air.
3. Low-level waste discharged into the ground.

No satisfactory method of disposal for nuclear wastes had been found. Should the waste be buried in deep underground caves? Should it remain at the plants that produced it or are there isolated sites where waste can be shipped? Will storage facilities leak and contaminate the environment? If waste is shipped or not closely guarded, can it be diverted to illegal purposes?

The Department Of Defense, DOD, has long argued that the Atomic Energy Acts of 1946 and 1954, which gave it sweeping powers of self regulation, should take precedence over environmental statutes such as, the Resource Conservation and Recovery Act, and Superfund Laws. In the last seven years, Congress has enacted legislation specifically requiring The Department of Energy to meet Standards set by these laws. Although the agency has agreed to comply in some instances, the Reagan Administration threw many obstacles in the way of enforcement, including policies that prevented the Environmental Protection Agency and states from suing DOE. The unsafe practices continue.

States such as South Carolina and Washington, home of the major plants producing bomb grade materials, are trying to force DOE to comply with the law. These states are afraid that if nuclear weapons production is cut back, they will have to cleanup the waste, which is lethal in even minute quantities, and will remain dangerous for thousands and even millions of year.

Chapter 16: Proposed Burial Sites for Nuclear Waste

Shallow Graves, Landfills, Deep Wells, and Grout Injection

In the 1940s and 1950s, most military and civilian nuclear wastes were buried in shallow graves in the ground on government property. Before 1970, the government packed low level waste into 55 gallon drums and dumped it at "sea." In the 1960s, government restricted its disposal sites to its own waste location, mostly generated by the weapons program. Since siting and construction standards for these early waste disposal sites were less stringent than they are today, and since some of the waste buried at the sites was quite dangerous, several of the sites threatened to harm public health or the environment and needed to be cleaned up.

Shallow land burial was the system currently employed for temporary storage of commercial and government solid LLW in the United States. Shallow burial which calls for waste to be deposited in trenches and capped with soil or concrete when full has long been practiced in the United States and elsewhere. It was first tried by the United States during World War II.

In working to develop the atomic bomb, wastes was dumped into trenches two feet deep at Palls Forest, a park preserve just twenty miles outside the city of Chicago, and later covered with concrete. In the years since, the waste have contaminated nearby underground and well water. They have also spread high concentrations of uranium and plutonium to adjoining meadows. Today the dump site is marked with a stone tablet that warns visitors; Caution, Do Not Dig, but Palls Forest continues to be a popular spot for hikers, picnickers, and campers.

Today the trenches are lined with concrete and contain special backfield materials. The adjacent ground is fitted with drainage systems to prevent rainwater from eroding their sides and tops. Their locations are chosen according to scientific guidelines and the waste that go into them are often locked in steel containers. What is the significance of the containers being locked? It is likely that shallow burial or a similar burial system in mined cavities will be the choice of many sites.

In the 1960s civilian low level wastes from nuclear power plants had been put into landfills at sites chosen by state governments. The waste was packed in containers and they were placed in large shallow trenches in the ground. Filled trenches were capped with clay or some other materials that discouraged rainwater from percolating through the trenches. Then the entire site was graded to control drainage and erosion, but nothing was put in the bottom or the sides of the trenches.

Government LLW was stored on selected federal lands and commercial LLW was buried at power plant sites or at commercially operated dumps that had been established for that purpose.

In ground percolation, the liquid LLW was directed to open ponds or underground cribs. It was then allowed to seep through soil, sand, and gravel to underground water sources at a depth of about 100 meters. This method was employed in the early years of the nuclear era when scientists believed that the waste, with their weak and usually short-lived radiation, would be rendered harmless when diluted in the vast expanses of "underground water." It is now known that the stronger and more durable LLW can pollute the underground water and the surrounding subterranean rock and soil formations. Ground percolation is no longer used in the United States or in a number of foreign countries, and is not seen as a feasible disposal system for the future.

The idea behind deep well injection was to mimic nature's way of storing petroleum, gas, and water, all of which have been locked in natural traps within the earth for millions of years. Bore holes or drill to fissures and caverns deep within the earth's crust and the liquid LLW was then sent downward to occupy those spaces. It was not successful.

Through the years, much LLW has been stored at power plants and reprocessing sites. The geologic strata beneath many of the installations have proved unsuitable because they are without fissures and caverns adequate for the process. Consequently, deep well injection is little used today and it should be ignored in the future.

Grout injection is similar to deep well injection, but involves an extra step. When the LLW liquid is injected into the ground, it is accompanied by grout, which is a plaster like substance. Once underground, the grout is suppose to solidify and lock in the waste's radiation. Grout injection held the promise of being an effective system for the state. Commercial LLW was buried at power plant sites or at commercially operated dumps that had been established for that purpose, but had the results been tested?

Alternative technologies included placing the waste in underground and above ground vaults, in concrete bunkers mounted over the earth, or burying the wastes more deeply. These alternatives were more expensive than shallow burial in trenches and not any safer.

Storage in Double-Walled Tanks

In the early 1950s double-wall tanks came into use as an extra safe guard against leaks caused by corrosion or other breaks in the tanks. Double-wall tanks were installed at the government facilities at Savannah River in North Carolina and Idaho Falls, Idaho. The Savannah liquids were neutralized with sodium hydroxide and stored in carbon steel tanks. Idaho Falls did not at first neutralize it wastes but deposited them directly into stainless steel tanks.

One of the Savannah River tanks began to leak in 1960. About 100 gallons of HLW escaped into the ground and contaminated a nearby source of underground water. The government has been criticized for these and other leaks and it has responded by claiming that only a mere fraction, less than 1 percent, of all its stored liquid waste had escaped via leakage.

In the years that followed, adding to the problems caused by Hanford's emissions of radioactive waste into the atmosphere, some 450,000 gallons of high level liquid waste seeped from 20 of the plant's 149 tanks. As of 1980, there were no reports of serious underground water pollution caused by the leaks, but what does serious pollution caused by leaks mean?

Nuclear Wastes Pumped Into Ground

Another choice was to bury the waste in a deep salt deposit near Lyon, Kansas, but was abandoned in the early 1970s because past drilling in the area for gas and oil had created a risk that groundwater might percolate through the site and become contaminated. Other burial sites were investigated, but none were selected by the time Congress passed the Nuclear Waste Policy Act in 1982. This legislation directed the Department of Energy to develop two deep burial sites for HLW and implied that one should be located in the West and in the East. It also set off a year of political wrangling since no state wanted to host such a permanent disposal facility.

Beginning in the early 1960s, LLW and HLW were pumped directly into the earth at three separate sites in Russia, Tomsk, Krasnoyarsk, and Dimitrovgrad. The well in which these unshielded wastes were injected were only 650 to 4,600 feet deep much shallower than the seven mile-deep nuclear burial option considered by some Western nations. Russian scientists in 1994 acknowledged that about half of all the radioactive materials produced in Russia had been injected at these sites, each of which is located near a major river. In 1995 scientists were still evaluating where the nuclear wastes were located, how they might move underground, and what dangers they might pose to the environment and to human health.

Waste had been dumped by the former Soviet Union extensively in the world's oceans. Not only the former Soviet Union, but also various Western nations have dumped nuclear waste in the oceans. Until 1970, the United States was among them. In 1993 Russia admitted they had scuttled decommissioned nuclear submarines as well as dumped tens of thousands of containers of HLW and LLW in waters. Russian spokesmen indicated, that they did not wish to continue ocean dumping, but that with few other options they might be forced to do so.

Solidify Nuclear Waste Pumped Into Ground

The government called for plants to solidify the wastes, a process that was to end the danger of leakage and also reduce the volume of waste. In response, Hanford and Savannah River evaporated the water content of their neutralized waste. Left behind in the tanks were a reduced moisture of liquid and slugs. At the Idaho Falls plant, where waste had not been neutralized, it was chemically treated and turned into calcine, a dry granular material. The calcine was then transferred to concrete-lined stainless steel bins underground chambers.

The government began giving greater attention to improving "temporary" storage facilities. Twenty-seven new tanks were constructed at Savannah River and twenty at Hanford. The new construction was intended to eliminate the single-wall tanks at both sites by transferring their liquid content to the new tanks. There was no problem with the transfer, but then a difficulty arose that stalled matters for years to come.

Remaining in the old tanks were the slugs. It had to be dissolved and removed by the use of nitric acid. The acid hastened the corrosion of the old tanks and threatened to release even more waste into the ground.

The highly radioactive portions were to be solidified and held for later permanent storage. The portions with lower radioactivity were to be injected into the earth along with the plaster-like substance, grout, which would then solidify and lock in their radiation they hoped.

Similar work was being done at the civilian reprocessing plant at West Valley, New York. The plant was being decontaminated and wastes were being cleaned up and removed from its various facilities. A smelter would then be installed at the site. The smelter would solidify the high-level-radioactive-waste and convert it into a glass-like form. The decontamination work was to be done under a joint federal/state plan.

Years of lax standards at United States Weapons Facilities allowed a great deal of contamination to endanger workers and the public, created expensive cleanup problems, and an unnecessarily large amount of hard-to-handle radioactive waste. Two facilities with especially bad records of contamination are the bomb-making plant at Rocky, Flats, Colorado, and the Hanford, Washington, Plutonium works.

Reprocessing Spent Fuels

Constructed in the 1960s, West Valley was the first civilian installation for the reprocessing of spent fuels. It was to neutralize its liquid waste and placed it underground in steel tanks. The tanks boasted an extra safety feature. Each had a giant "saucer" fitted beneath it to catch any leakage, a technique that was later used at Hanford. Throughout the six years of is operation, West Valley stored about 60,000 gallons of HLW. The plant was closed in the early 1970s because drainage problems caused some of its storage facilities to overflow and contaminate the surrounding area during rainstorms. Apparently it was not considered that water that drain through the earth would also fill the saucers causing them to overflow.

By the 1980s, the leaks and the mounting accumulation of solid High Level Waste at both defense and civilian nuclear plants, left no doubt that steps must be taken to establish a definite system for permanent storage.

Nuclear Waste Buried In Ocean Floor and Space

As its name indicates, sub-seabed emplacement proposes that the waste be deposited below the ocean floor in any of three locations. The first two were deep sea trenches and mid-oceanic fractures zones, areas where the ocean floor is split open. The third, which many scientists thought the best, was deep-sea layers of clay at the center of stable areas of the ocean floor. However it was later rejected. One problem, the burial was to be carried out in international waters. Another, the process would require transporting the waste canisters over distances. Turbulent seas and hazardous weather could sink the transport ships and result in the contamination of such areas as coastal waters and commercial fishing grounds with radioactivity.

Instead of burying the waste canisters deep beneath the ocean floor, a new plan was to fire them into deep space and allow them to float there. Or it would store them aboard rockets and then launch them toward the sun. Danger of launch of any space vehicle and cluttering space with any kind of garbage should have been unthinkable for any scientist.

Nuclear Waste Stored In Ice Sheets

This idea was based on the fact that HLW emits great heat. Waste canisters would be placed on the vast ice sheets that cover the island of Greenland and the continent of Antarctica. The heat radiating from the containers would melt the ice enabling them to sink deep into it. When the ice cooled again, it would solidify around the canister and lock in their radioactivity.

Department Of Energy first called passive slow decent, would insert the canisters in shallow holes and then allow them to sink until they reached the bottom of the ice sheet. The second, known as the anchor concept, was similar to passive decent, except that a cable attached to each canister would permit it to sink no father than a designated level and would enable it to be retrievable at a later date in the event of an emergency. The third method was known as surface

259

emplacement. It involves constructing large storage units on the surface of the ice and filling them with the waste canisters. The heat radiated would cause the units to melt their way slowly to the bottom of the sheet. Was the rate of melting to be regulated and was the melting of ice under the storage units controlled for level displacement?

The burial would be made in remote and desolate regions. Antarctica is uninhabited by humans, and thus the waste would be in almost total isolation. The burial would be deep because the ice sheets are amazingly thick, the thickness of some ranging up to 10,000 feet or more, but how deep would they sink? They are formations that, unlike the earth, remain stable for long periods, insuring that the burial would not be disturbed by such upheavals as earthquakes.

Hazards of transporting the wastes over great distances to the disposal sites were later considered. The transport ships would be sailing into the world's worst weather with bitter polar cold, and the long periods of polar darkness, making the burial difficult and dangerous. The United States along with a number of other nations, had signed the Antarctic Treaty of 1959 that inhibited the disposal of nuclear waste on the continent at the bottom of the world.

Nuclear Waste Buried In Cavities on Islands and In Deep Mines

Like the process of burial in mined cavities, island disposal was considered a system of geologic burial. The wastes were to be interred not in masses of ice, but in natural or man-made caverns lying beneath the earth's surface. This particular type of geologic burial recommended that the burial took place on uninhabited islands far from civilization.

Because the islands were uninhabited, the danger to health from radioactivity leaking into the atmosphere would be nil. Any leakage into the island's underground water would also do no one any harm, but the Department Of Energy saw two disadvantages. The danger involved in shipping the wastes to distant burial sites. Second, that many islands were prone to intense seismic and volcanic disturbances. Either could rupture the canisters and discharge enough radioactivity into the atmosphere to threaten distant inhabited locals.

Burial in deep mine cavities proposed a complex network of tunnels be constructed approximately six miles beneath the earth's surface. The tunnel floors would then be fitted with shallow holes into which the waste canisters would be inserted to keep them securely in place. To insure that they would not bump against each other during any shifts of the surrounding geologic formations, the spaces between them would be covered with thick layers of dirt. Once filled to capacity with containers, the tunnels would be completely sealed off.

The Department Of Energy subjected the plan to serious research for a time, but finally turned it down. Drilling of holes to a depth of six miles and then the construction of tunnels at that dept were impossible. No one knew how a rod at that depth would withstand any geologic upheavals or the intense heat emitted by the HLW. Although the concept of burial in deep-mined cavities was rejected, it did point the way to the system that Congress finally decided should be used for the permanent storage of HLW. That system was specified in the Hanford Drama.

Burying Nuclear Waste Deep Underground in Ice

Like ice sheet emplacement, it was based on logic burial. Based on the fact that HLW gives off great heat, it would require drilling holes down to fissures and cavities deep within the earth and then placing the wastes there. Once the water in the wastes had "evaporated" and converted into solids, they would emit such an intense heat that the surrounding rock formation would melt. Melting would create a sphere of molten material that would eventually begin to cool and solidify. The results would be a solid mass of waste and rock in which the radioactive elements were trapped.

The rock would need an excessively long time to solidify and seal in the radioactivity, sometimes a thousand years or more. In that time, too many things could happen below the earth's surface that might render the method unsafe.

Nuclear Low Level Waste Dumped In Sea

Beginning in the late 1940s solid LLW were packaged in 55 gallon and 80 gallon steel drums and dumped into the sea from "ships and aircraft" operating off both the Atlantic and Pacific coasts. The drums were lined to 55 percent of capacity with concrete. The concrete served two purposes. A protective measure against radiation leakage, and weighting the drums so they would sink to a depth of some 12,000 feet. They were to be dropped in designated areas and often those points were ignored by the waste-carrying ships and planes when bad weather prevented an aircraft from reaching the designated spot before running low on fuel. When aircraft or ship crews simply wanted to save time, they would dump the waste and get home at an earlier hour.

The captains of commercial ships under contract to the government often returned to shore quickly so that they would not have to pay their crews overtime wages. In such cases, the drums would be short dumped or dropped in any convenient spot, regardless of the depth or the proximity to a fishing ground.

Many of the drums were not sufficiently weighted and failed to sink. They floated on the water surface and when sighted, were not picked up for proper repackaging, but were split open and sunk by naval gunfire. There radioactive contents spilled into the ocean to pollute the waters and endanger the surrounding sea life. It was believed that the oceans were so vast that the waste would be dispersed and diluted enough to make it harmless. But most of it was believed to stick to the sea bottom and lie there as a constant threat to the surrounding marine life, and a treat to anyone who ate fish.

Great Britain generated havoc in the Irish Sea by discharging tons of HLW, deadly plutonium, ruthenium, and cesium-137 into its waters. News reports say the British government's power plant and reprocessing center at Sellafield has leaked a quarter-ton of plutonium into the Irish Sea over a thirty-five year period.

Releasing Nuclear Waste into Atmosphere

In early 1986, the government nuclear plant at Hanford, Washington, earned headlines across the nation when an environmental group reported that the installation had been disposing of its waste materials for years by releasing them into the "atmosphere." The report on secret United States government documents it had secured under the recently enacted Freedom of Information Act. The documents revealed that the plant had loosed great amounts of such dangerous radioactive wastes as iodine, ruthenium, and cesium in clouds of gases that had rolled out for miles from the plant. Did fallout from the clouds of iodine and other radioactive wastes harm the surrounding area?

People in the small farming town of Mesa, which lies about ten miles east of the Hanford installation were convinced that they suffered much harm. One farmer, his wife, and three of their daughters all took medication for thyroid problems. Another resident reported that his father had colon cancer, and his mother skin cancer and his two sisters had their lower colons removed. The man says that he himself is sterile and has only 90 percent of his lung capacity.

Just outside Mesa was a rural area that had won the ugly nickname, the "death mile." Of the 108 people who lived there while the clouds of gases were rolling in from the Hanford chimneys, 24 became ill or died of cancer since the 1960s. Also feared are the "200 billion gallons" of radioactive waste water that the Hanford plant poured into ponds and pits over the years. The concern centered on whether or not the waste water, which resulted from the reprocessing of spent fuels, had filtered down to contaminate the area's underground water supply. The Time Magazine estimated that the amount of Hanford's dumped water was sufficient to create a lake some 40 feet deep and a size to cover New York City's Manhattan.

Government LLW is stored on selected federal lands and shallow burial was the system used. Commercial LLW are buried at power plant sites or at commercially operated dumps that have been established for that express purpose.

Permanent storage was set up in the 1970s at Hanford, Washington, Barnwell, South Carolina, West Valley, New York, Maxey Blats, Kentucky, and Sheffield, Illinois. What plans are under way for the permanent storage of the waste now? The answer begins with two pieces of legislation passed by the United States Congress during the early 1990s.

Waste Dumped On Indian Reservation and In The Columbia River

The Hanford Reservation is one of the oldest and largest of the nation's nuclear weapons facilities. It consisted of 560 square miles which is half the size of the state of Rhode Island. It produced plutonium for nuclear weapons from 1943 until 1998 and was home for nearly two thirds of the weapons program's total volume of solid and liquid hazardous and radioactive waste. That included 65 million gallons of high level reprocessing waste stored in 177 underground tanks.

Pressed for time during the war, Hanford's engineers decided simply to dump LLW on the Indian reservation and into the Columbia River, expecting the environment to dilute them. HLW was stored temporarily and the engineers expected they would figure out what to do with the waste after the war. Radioactivity from Hanford was detected 200 miles downstream in the Columbia River that ran by Hanford and was fed by groundwater beneath the reservation.

About 440 billion gallons of other waste leached into the ground at Hanford over the years. One hundred square miles or more groundwater was contaminated with radioactive and hazardous chemicals. Cleaning it up is expected to cost more than $57 billion and to take longer than thirty years, if it can be cleaned up.

Nuclear wastes have been dumped at more than 1,400 locations on the Hanford Reservation. The wastes include enough plutonium to build two dozen nuclear weapons. Most of the volume of what have been dumped consisted of LLW ranging from workers contaminated clothing to contaminated heavy equipment, such as bulldozers. It is even rumored that workers on the reservation once laid track into a pit into which an irradiated locomotive was driven and buried.

The waste records at Hanford were haphazard for the early years, so nobody knows exactly what's there, or even all the places where waste has been buried.

DOE contractors were to move hazardous materials from all over the reservation into a smaller zone, the 200 Area where tanks of high-level waste were supposed to be buried. That section was so badly contaminated that it may never be cleaned up. The 200 Area was to be a temporary storage site for the transported hazardous waste. A permanent solution had not been determined, but concentrating waste in the 200 Area was to allow cleanup to proceed more quickly on the rest of the reservation.

In May 1993, after years of argument, government officials, environmentalists, scientific experts, and representatives of Native American tribes and other local interests agreed on an overall cleanup strategy for Hanford.

As of 1995, high level waste remained in tanks on the reservation. As early as 1948, the Atomic Energy Commission's Committee on Nuclear Methods permitted pits designed for LLW to seep into the ground from corrosion-prone single-wall tanks for HLW.

During Hanford's peak years, nine nuclear reactors operated at the reservation. Waste disposal and storage continued to be haphazard. An explosion shot hundreds of contaminated glass shards into a 64 year old worker named Harold McCluskey in 1976. Miraculously, he survived and lived to be 76 years of age, but what does "lived" mean?

Secretary Herringbone said a 1991 report by Congress's Office of Technology Assessment found evidence that "air, groundwater, surface water, sediments and soil, as well as vegetation and wildlife, had been contaminated at most of the Department of Energy nuclear weapons sites."

The worse problem in the 200 Area at Hanford is the leakage from the underground tanks. Every year another two of three of the huge steel tanks which were not designed for long term use, begin to leak into the surrounding soil. The leakage has amounted to more than 1 million gallons in addition to the unknown quantity of liquid that in years past was routinely siphoned off from the tanks and dumped on the ground. By mid-1993, 68 of the 177 tanks were on the DOE's list of "assumed leakers."

The scariest aspect of the tanks was not leakage, it was the threat that gases bubbling up in the tanks might ignite. An explosion of similar waste tanks in the Soviet Union in the 1950s spewed radioactive contamination over thousands of square miles on the southern Ural Mountains. In 1993 a waste tank holding a brew similar to what was kept at Hanford exploded at the Russian reprocessing plant at Tomsk-7.

In 1993 an elaborate $30 million mixing pump was installed in a tank whose type considered the most dangerous. The theory behind the mixer was to bleed the tank's gases off steadily, rather than letting large bubbles build up to be emitted in dangerous "burps." Technicians timed the installation after the tank had burped thousands of cubic feet of hydrogen.

After the mixing pump was installed, a Hanford worker was contaminated with radioactive wastes after lowering a "rock" on a rope into one of the tanks to check if one of the pipes was blocked.

Early plans called for the waste in the tanks to be separated. The most hazardous part of the waste was to be chosen then shipped to the permanent HLW storage. The remainder was to be mixed into wet cement, and the wet cement poured into giant vaults and left at Hanford. The public rebelled at the idea of the concrete vaults. Eventually, it was decided to vitrified all of the wastes in the tanks, even though this would create about 38,000 glass logs, 10 feet long by 2 feet wide. Vitrify is mixing glass with the waste using heat.

Three of these dingy gray hulks sprawl about a thousand feet long with walls of reinforced concrete up to eight feet thick. Tear these down and where do you put that rubble? Some suggested to fill them with LLW and cover everything with dirt.

By the early 1990s the DOE was spending more than $1 billion each year at Hanford. That is more money than was spent when the plant was actually making weapons. Nearly all of the money was being spent on planning rather than on actual cleanup operations.

At the end of 1993, plans were drawn up calling for building and operating a larger-capacity vitrification plant, keeping the vitrified remnants at Hanford. The whole process was to be completed by 2028, assuming that all of the project's technical hurdles were overcome.

When you read about the many bad ways the trained minds thought of storing nuclear waste knowing it could be radioactive for years, you wonder what was the thinking processes used.

Andrew D. Anderson

Chapter 17: Nuclear Waste Leaking and Health Problems

One Site, Over One Millions Gallons Leaked - Over $3 Billion Spent - Could Not Stop Leaks

The Waste Isolation Pilot Plant in Southeastern New Mexico near Carlsbad was for defense waste. The construction of some burial rooms has already been completed. It is below the surface in the salt beds of the Salado Formation, and was intended to house up to 6.25 million cubit feet of transuranic waste and expected to be ready for operation before 2000.

The federal government was also building a repository for the storage of most of the liquid wastes generated in the reprocessing of spent fuels for defense purposes.

The Department Of Energy intended to bury its transuranic waste permanently at the Waste Isolation Pilot Project (WIPP). Some radioisotopes will remain extremely dangerous for anyone to ingest during the next 10,000 to 100,000 years of more.

The WIPP is a cavernous burial site dug out of salt deposits 2,150 feet below the ground. It could hold as much as 220 million cubic feet, far more than the volume of transuranic wastes in storage. The temptation to bury additional waste may be difficult to resists. By 1993, $1.5 billion had been spent on WIPP and an additional $1.5 billion was spent by the year 2000 totaling $3 billion.

Many environmentalists believe that WIPP is fatally flawed. The burial chambers are dug out of a salt deposit that is sandwiched between a pressurized brine reservoir "below" and a freshwater aquifer that feeds the Pecos River above. Water seeps through the salt walls of the burial chambers, is highly corrosive, and this briny liquid could corrode the steel drums in which transuranic wastes are stored. Worst, the wastes give off gases that could build up pressure after the site is sealed, open up cracks in the salt deposits, and force contaminated material to the aquifer above.

The presence of the water caused the DOE to plan on loading the repository to just 2 to 3 percent of its capacity of about 1 million canisters. Each canister contains 55 gallons of wastes. The planning, studies, and primary work was flawed. Will this happen at Yucca Mountain?

With so much money already spent on WIPP and no place else to put the wastes, the DOE was strongly motivated to go forward.

Increased Rates of Cancer and Birth Defects

A DOE report dated 1992 stated that the Hanford cleanup was failing and efforts to stabilize the tanks were not keeping up with their deterioration. It was estimated that cleaning up Hanford would cost more than $57 billion and take longer than thirty years if it could be cleaned up.

In 1992, DOE in Portsmouth Gaseous Diffusion Plant in Ohio turned up 578 health and safety violations, according to an aide to Senator Glenn. A 1993 DOE study showed that the department was paying its contractors one third more to clean up nuclear sites than private industry pays for similar projects. This expensive work was usually completed late, and more than half of the contracts studied generated cost overruns.

The Hanford inventory included 53 million gallons of waste from plutonium processing stored in underground tanks, nearly 2,300 tons of spent fuel, four and a half tons of plutonium, 25 million cubic feet of solid waste, and 38 billion cubic feet of contaminated soil and groundwater.

In storage pools at the nation's most lethal single source of radiation was 1,936 steel cylinders containing cesium and strontium covered by 13 feet of water. When a technician switched off the lights, radiation from the cylinders puts on a light show of royal blue.

Water taken from the nearby Columbia River to cool reactors was returned to the river with a burden of radioactive sodium, zinc, arsenic, and other elements. Later, waste stored in underground tanks leaked into the soil, and 45 billion gallons of contaminated liquids were dumped on site, some near leaking tanks. Thus contaminated plumes were created underground, threatening the Columbia River and the press began reporting claims of increasing rates of cancer and birth defects in people and animals on farm areas near Hanford.

Technicians nonchalantly recorded that the radioactive gas was spreading above ground farther than anticipated. They just enlarged their sampling circles to 25, 50, 100, 150 miles, all the way to Spokane and Walla, but did not reduce the emissions.

Contamination was found in desert flowers decorating the desk of an official. Concern mounted, but Hanford had plutonium production quotas to meet. Special silver filters finally stopped 99 percent of iodine emissions.

No Records of Amounts of Nuclear Waste Stored

Makhijani studied government records and concluded that officials were guessing. "They were throwing darts." He sent a critique to DOE and they analyzed it for two years and finally agreed with him, admitting there was ten times more radioactive plutonium-contaminated waste buried in pits throughout the nuclear weapons establishment than they thought. "There is a ton of plutonium in Idaho alone," said Makhijani. "Some of it is leaching through the soil and threatening the Snake River aquifer."

Marcus Page wants to abolish nuclear weapons and power plants because they create more waste. There is no safe way to store it, so it is irresponsible to generate radioactive materials that last for hundreds of generations.

A Serious Nuclear Waste Storage Problem

There are no plans for managing nuclear waste and the waste we have mentioned is only the beginning. As more and more weapons, plants, and nuclear power reactors come to the end of their useful lives, decommissioning and cleaning up the sites will generate still more waste.

How much cleanup the American people are willing to pay for will ultimately be decided in the political arena. If they do not buy it, what is the alternative? Will decontaminated nuclear sites be clean enough that they can be released for unrestricted use, building sites for homes or schools? Which contaminated sites should be cleaned up first? Those presenting the most immediate dangers to human health or the environment, those that are easiest and cheapest to clean up quickly, or those that have the most typical problems so that lessons can be learned for future cleanups?

Dismantling of a nuclear power plant not only makes what to do with the spent fuel a more urgent concern, but also generates a huge pile of low-level waste. More waste than was generated while the plant was running. With so many plants expected to be decommissioned in the future, a mountain of low level waste lies ahead. At present we have no place to dispose of any of it.

The 1980 Act was Congress's response to an approaching low level waste crisis. One by one, disposal sites around the country were closing. By the end of the 1970s only three remained open. The site of Barnwell, South Carolina, received 85 percent of the nation's sites containing low level nuclear waste from Three Mile Island. In 1979, it announced plans to cut in half the amount it would accept. The nation's other two sites, in Nevada and Washington, also planned to institute restrictions.

The 1980 legislation made each state responsible for finding some way to dispose of the waste generated within its borders. By

1995, not a single new disposal site had been built. A sampling of the local and environmental opposition to specific sites and the political wrangling among the states demonstrated how this impasse developed.

1. In 1992, voters in Boyd County, Nebraska, overwhelmingly rejected a proposal to a low level disposal site in their county. The project would have received waste from Nebraska, Arkansas, Louisiana, Kansas, and Oklahoma.

2. In 1993, South Carolina threatened to cancel New York State's contract for dumping wastes at Barnwell, charging that New York wasn't serious about finding an alternate disposal site and emphasizing the residents of Barnwell wanted their site close. In response, New York began to consider reopening a low level site at Ashford, New York, that had been closed in 1975 when waste in the landfill overflowed.

3. Toward the end of 1993, citing unresolved environmental issues, the Clinton Administration put on hold a transfer of federal land to California for a disposal site at Ward Valley in the eastern Mojave. Transuranic waste was packed in 55 gallon steel drums at the Savannah River site and stored there. Once all the space was used, the slab was covered with earth. In 1994, as political pressures against the project mounted, a spokesman for the company that would run the project said, from a national perspective, if we cannot put a disposal site here, we cannot put one anywhere.

In the 1990s, about half of the volume of civilian low level nuclear wastes generated each year came from nuclear power plants. One third of the volume came from industry, which used relatively small quantities of radioactive materials in enterprises ranging from biotechnology research to quality control operations. The rest, a comparatively small amount, came from a great many individual sources including hospitals, clinics, universities, and other research centers.

The 1980 Act permitted the older disposal sites to refuse to accept waste from outside their regions after January 1, 1993 whether or not new disposal sites were open by then. No new sites opened. By 1993 the Nevada site was closed and the Washington site was refusing to accept waste from outside its region, leaving only the Barnwell site to accept all the rest of the nation's nuclear wastes. Barnwell slapped a $220 per cubic foot surcharge on all wastes from outside its region. It declared it would stop accepting any waste from outside its region in 1995, and announced that it would shut down altogether in 1996. Low-level nuclear wastes had piled up in many sites around the country, as users with no place to go had improvised storage sites for their wastes.

High-level nuclear wastes presented greater problems. Although spent fuel from nuclear power plants accounted for only about 1 percent of the volume of all nuclear waste. It contained about 95 percent of all the radioactivity of military and civilian waste combined and called for very careful handling.

Storage Waste Sites Reducing

The Berkeley Pit Mine near Butte, Montana, is the nation's largest federally designated hazardous waste site. More than a mile across and 1,800 feet deep, the hole contains millions of cubic yards of mine tailings that feed into a 125-mile stretch of river that has become contaminated.

The EPA reported in 1993 that the states with the highest level of chemicals released into the air, land, and water each year by industry were Texas, Louisiana, Ohio, Tennessee, Indiana, Illinois, Michigan, Pennsylvania, Florida, and Kansas. Because of the problems in finding disposal sites, the United States is sending larger and larger amounts of toxic waste out of the country. Mexico, Central, and South America have become preferred spots for sludge and incinerator ash. Occasionally, toxic waste is mislabeled nontoxic when it arrives in South American countries. Africa had been a favorite location and in 1988, African countries signed agreements that restricted importation of dangerous materials.

In 1985, 837 of the nation's 1,538 land disposal facilities for hazardous waste were required to close because they failed to meet RCRA requirements. Those units then came under mandate for cleanup, but because of lack of funds, follow-up inspections and enforcement actions lagged considerably.

The EPA has the authority under the RCRA to require businesses with hazardous waste operations to take corrective actions to clean up the waste that has been released into the environment. An estimated 3,400 facilities out of about 4,300 in the RCRA universe are suspected of contaminating the environment. Corrective action is a slow process.

About one fifth of the sites for cleanup by Superfund were municipal landfills. Some were federal sites, including those for the disposal of nuclear materials from bombs or United States Air Force bases that did not properly dispose of fuels and other dangerous materials.

In 1990, an EPA report Reducing Risk for All Communities, concluded that racial minorities and low-income people boar a disproportionate burden of environmental risk. Racial minorities and low income people were exposed to lead, air pollutants, hazardous waste facilities, contaminated fish, and agricultural pesticides in far greater frequencies than the general population.

In a Chicago laboratory, Italian physicist Enrico Fermi assembled enough uranium to cause a nuclear fission reaction. His experiment also produced a small packet of radioactive waste materials that will remain dangerous for hundreds of thousands of years. That site or the original waste lies buried under a foot of concrete and two feet of dirt on a hillside in Illinois.

In the 1950s and 1960s while spending only a few hundred million dollars to research storage and disposal processes, the nation spent billions of dollars to produce nuclear weapons and to commercialize nuclear power.

More Storage Waste Sites Closing

In 1992, controversy arose when a number of nations demanded that Japanese ships remain out of their territorial waters during the process of shipping the largest sealift, over one ton of plutonium, in history. South Africa, Chile, Malaysia, and Indonesia told the Japanese that its transport ship carrying roughly a ton of plutonium from reprocessing centers in Europe to civilian reactors in Japan would be barred from passing through their territories. Also radioactive nuclear materials have been smuggled from the former Soviet Union into Western Europe and the Russian government failed to protect their own radioactive materials.

According to the DOE, which manages nuclear waste disposal in the United States, there are 20,000 metric tons of nuclear wastes in what is called "spent fuel pools" at the 109 operating and 20 closed nuclear energy plants around the country. By the year 2000, the total waste is expected to reach 50,000 metric tons, according to the Nuclear Regulatory Commission which licenses power plants. Almost all of these plants will reach their capacity for storage before the end of the decade.

Many nuclear plants are shutting down well ahead of schedule because of skyrocketing maintenance and repair costs. NRC licenses power plants to operate for 40 years but they do not last that long. The average life of the more than 20 reactors that have been shut down has been around 13 years. Was this money well spent?

When they are closed, the nuclear waste and the radioactive equipment stay on the premises. There is no place to put it. As a result, every nuclear power plant in the United States has become a temporary nuclear waste disposal site. Those that close become mausoleums, largely untouched while they wait to be decommissioned or dismantled when a repository eventually opens. The states are under federal mandate to find storage for the radioactive waste they generated which is currently shipped to storage sites in South Carolina and Washington.

The DOE is responsible for cleanup and waste management at 15 major contaminated facilities and more than 100 smaller locations in 34 states and territories. Cleaning up these sites is an enormous task. DOE's most recent estimate is that the cleanup will cost at least $300 billion, perhaps as much as $1 trillion and take more than 30 years to complete if ever.

The DOE was also authorized to develop a facility to receive and temporarily store waste until a second repository is built. In March 1987, the DOE proposed developing a facility in Tennessee for temporary waste storage to begin accepting waste in 1998. Congress authorized the plan, but it would not go into effect until the Nuclear Regulatory Commission had authorized the construction of the Yucca Mountain repository.

The federal government plans to take apart as many as 15,000 warheads and store them at a weapons plant, Pantex, near Amarillo, Texas. The government must then decontaminate buildings used at those facilities, dispose of millions of gallons of boiling radioactive water, and decontaminate hundreds of square miles of desert at the Nevada test site.

Additional Problems for America

Invisible radiation intense enough to kill emanates from 11.9 tons of spent nuclear reactor fuel, confined in steel cylinders behind glass walls at a government facility in Idaho. The United States is seeking a permanent repository for such dangerous waste which is just part of the nuclear disposal problem. The Bush administration's choice for high level nuclear waste storage, Yucca Mountain, has been studied for over 20 years by the Department of Energy at a cost of $4 billion dollars.

Deep underground, armed guards patrol plutonium stored at Rocky Flats, a former weapons plant that closed in 1999. It had been designated a Superfund site three years earlier. Now in 2002, the sprawling complex is being dismantled and its plutonium slated to be moved.

A long deferred cleanup is now under way at 144 of the nation's nuclear facilities, which encompass an acreage equivalent to Rhode Island and Delaware combined. Many smaller sites, the easy ones, have been cleaned, but the big challenges remain. What's to be done with 52,000 tons of dangerously radioactive spent fuel from commercial and defense nuclear reactors? What is the United States to do with 91 million gallons of high-level waste left over from plutonium processing, scores of tons of plutonium, more than half a million tons of depleted radium, millions of cubic feet of contaminated tools, metal scraps, clothing, oils, solvents and other waste? There are also 265 million tons of tailings from milling uranium ore. Less than half of that amount stabilized littering landscapes.

For an idea of scale, load those tailings into railroad hopper cars, then pour the 91 million gallons of waste into tank cars, and you would have a mythical train that would reach around the Equator and then some.

In addition to storing the waste, contaminated soil and groundwater must be treated and stabilized, nuclear reactors decommissioned, buildings demolished, some buried waste exhumed, sorted, and buried again because it was not buried right in the first place. The bill for all this will be staggering perhaps $400 billions over years, but "can it be done with any degree of efficiency?

Rocky Flats sits between Denver and Bolder and their galleries of critics religiously chronicled tainted groundwater, drums oozing waste, plutonium contaminated air ducts, pipes, and soil.

Plutonium's nasty habit of being pyrophorc, lightning spontaneously caused two major fires and myriad small ones, contributing to Rocky Flats reputation as one of the vilified weapons plants in the United States.

Rockwell, the plant's contractor, eventually plea bargained environmental crimes including acid spills and four other felonies and paid $18.5 million dollars in fines.

Storage casks at the Prairie Island nuclear power plant near Minneapolis have nine-inch-thick skins of carbon steel and hold 17.6 tons each of spent fuel assemblies. With the number of above ground casks limited to 17 by the state of Minneapolis, the plant will run out of storage space in 2007.

Officials from eight nuclear power facilities, including Prairie Island, have signed a lease with Goshute Indians to store spent fuel on their reservation in Skill Valley, Utah. We will get money for schools and a hospital says tribal chairman Leon Bear, who supports the plan. If approved by the Nuclear Regulatory Commission, the 100 acre site could hold up to 44,000 tons of spent fuel. Used fuel rods fresh from the reactor, intensely hot and radioactive, were stored at the plants in their metal cases on racks submerged in pools of water called swimming pools. Pumps circulated the water in the pools, dissipating the heat thrown off by the cooling fuel. The spent fuel was highly radioactive.

With nowhere else to put the waste, they were looking at a policy of at-reactor storage. But commercial nuclear reactor sites were selected with no more than 40 years of use in mind. To keep wastes at the plants indefinitely raised several issues. Many nuclear power plants were located beside rivers. The Palisades nuclear power station in Covert, Michigan, running out of room and its water-cooled storage, had been seeking permission to move some fuel into dry casks.

A nuclear power plant on the Mississippi River, near Minneapolis, would have shut down in 1995 unless it received permission to move some of its spent fuel into dry-cask storage. Pressured by local residents and environmentalists was applied and the state legislature balked at giving the plant permission.

The Energy Department is expected to move several tons of plutonium and weapons-grade uranium from a federal research laboratory in Alamos, New Mexico to Nevada in August 2002 because of security concerns. Several tons of highly enriched uranium and plutonium, that can be used to make an atomic bomb is kept in a Technical Area where critics said it could not be adequately protected. The area was built in the 1940s at the bottom of a steep canyon, where the high ground and an adjacent highway makes the site difficult to defend.

When you consider some of the places nuclear waste was scheduled to be storied and where cites were located, you wonder what was the criteria used to make those decisions.

Planning an Underground Nuclear Repository

This system called for the construction of what would be known as an Underground Repository. It was to be a complex network of tunnels in which the waste would be deeper than needed for the LLW, but not as deep as deep-mined burial.

Estimates held that the entire cost of establishing the facility and putting it in operation would cost between $21 and $42 billion. If state or Indian land was selected for use, both had the right to study the plans being made for the construction and operation of the installation and to challenge them and work for their change. They stipulated that the state and tribes be given financial assistance to cover the costs in their efforts.

These rights were granted to the Indian tribes because the American Indian Religious Freedom Act of 1978, which bars the federal government from interfering with the rights of Indians to practice their religious beliefs. It states that American Indians must always have access to their sacred grounds and objects. Because land and nature are basic to a Native American's religious worship, care was taken that neither was violated by the repository planning and construction.

It is interesting to note that all nuclear storage facilities are located or planned to be located on an Indian reservations because there is nothing there and nothing can be grown to eat.

The Environmental Protection Agency, which safeguards the nation's environment, was to establish standards that would protect the public from the hazards that accompany the storing of waste. Finally, the act established a special department, Office of Civilian Radioactive Waste Management (OCRWM) within the DOE.

The storage facility was to be below the ground surface at a depth of 1,000 feet in storage tunnels. The waste was to remain there for thousands of years in canisters filled with waste and sealed to prevent the escape of radiation. Until the repository was filled to capacity, it would be possible to retrieve the containers and haul them up to the surface if they were needed for some reason. When the repository reached capacity, it would be filled with dirt and sealed off. The underground workings were planned to spread over 1,500 acres.

A series of ramps and shafts would link the surface facilities to the underground operations. The principal ground-level building would be the waste handling plant. Here, the incoming waste would be received and given the final preparation necessary for burial.

Other ground-level facilities would include rail and truck unloading areas, warehouses, administration buildings, water and sewage treatment plants, a security office, and an area for holding the rock excavated in the construction or extension of the underground tunnels and chambers. For security, an area three miles wide would surround the storage site. This area would be controlled by the installation's security force.

The DOE set about looking for two sites, one that was needed immediately and one to be used at an unspecified date in the future. The department planned to have one serve the nuclear operations in the Eastern half of the nation, and the other for those in the West. It was felt that two widely separated locations would provide the best service by reducing the number of miles that the waste would have to travel en route to final storage.

Some of the things considered:

1.   Are the underground formations solid enough to seal in any radiation from stored waste?
2.   Are the underground formations stable or will they damage the tunnels and stored casks?
3.   Are they susceptible to such natural disasters as earthquakes or floods?
4.   Is there water nearby that will leak into the tunnels and cause them to collapse or corrode the waste canisters and permit radiation to escape?
5.   Will the location of the repository reduce or increase problems of safely transporting the wastes? How will it affect the costs of transportation?
6.   Will the site be too close to populated areas?

The DOE set the opening in the year 2002 at the earliest. Later, the date was again set to 2006. The department said it was postponing indefinitely the search for the site in the Eastern half of the nation and was concentrating its efforts on locals in the West. The quest later focused on three Western locations.

The department was to make certain that HLW are always transported in casks that were certified by the Nuclear Regulatory Commission. To be certified, the casks had to meet certain standards as to their strength and resistance to leakage and breakage by accident. Casks are the containers in which the wastes would be shipped to their final destination. Canisters are the containers in which they will be actually buried in.

They had studied three principal Western Locales, Deaf Smith County, Texas, Hanford, Washington, and Yucca Mountain, Nevada. Congress later told the DOE not to consider Deaf Smith County and Hanford.

In Texas the Deaf Smith County site lay in the midst of rich and valuable farmland. The construction of the repository tunnels would require that holes be drilled down through two major pools of underground water, one of which is the Ogallala Aquifer, a chief water source for the nation's Midwest.

In Washington, the Hanford site was finally rejected because it is near the Columbia River. Many local residents feared that radiation leakage might contaminate its water and it was too costly.

The department was to focus its studies on Nevada's Yucca Mountain. Yucca Mountain is a six-mile long ridge rising above barren desert land near the state's southwestern border. The mountain and the land around it are owned by the federal government and all repository studies were to be centered there. It would be the establishment and operation of a system for burying the wastes in deep underground tunnels that would be sealed off for thousands of years when filled to capacity.

Chernobyl Plant in Russia

In April 1986, the worlds worst nuclear power accident occurred with explosion and fire at the former Soviet Union's Chernobyl plant in Ukraine. The months that followed, 250 persons died. The release of radioactive material continued for 10 days, scattering fallout over the former USSR, European Countries, and the rest of the Northern Hemisphere.

The melt down that occurred hurled 7 tons of radioactive debris into the atmosphere. High winds carried the hazardous material as far away as Sweden and Great Britain to the West, and Italy and Turkey to the South. The Soviet government reported that the blast killed two people and hospitalized 197 with injuries. Thousands of others in the USSR and elsewhere were exposed to increased levels of radiation and an area of 2,500 square kilometers around the plant was rendered uninhabitable by the radioactivity.

In Turkey, homegrown tea remains unsafe to drink. Food grown in Central Sweden and Northern Italy is unfit to eat. Vast numbers of sheep in Scotland and reindeer in Scandinavia's Lapland are unmarketable because they have been contaminated with radiation.

With in three months of the explosion, twenty-eight people in the damaged region were reported dead of radiation poisoning and 300 were treated for serious radiation exposure. The next fifty years will bring 6,500 additional cancer deaths to the Soviet people living outside the Chernobyl area. Beyond the USSR's borders, an additional 10,400 deaths from cancer are predicted for Europe during that same period. Five hundred are predicted for Asia's, and fewer than 100 for Canada and the United States. Estimated for the total cancer deaths that can be expected in the next half century, 35 million of the Soviet Union, 34 million for Asia, and 49 million for Canada and the United States combined.

Also the Chernobyl disaster has revealed that the occurrence of lip, mouth, and other cancers has doubled among the residents of a nearby farm area. One Soviet newspaper has reported that half the children in the Narodichsky region of the Ukraine now suffer from an unusually high number of thyroid problems. Another paper has said that farm animals have birth defects like born without heads and limbs.

It usually takes hundreds of thousands of years for the radioactivity of nuclear waste to stop radiating. The original solution was to bury waste deep into the earth, but scientists now believe the deadly debris cannot be guaranteed to remain sealed off from the biosphere for hundreds of centuries. Despite the danger, storage remains the preferred option well into the twenty-first century. The total waste accumulation is 84,000 tons, many times more than in 1985.

Scientific expeditions were underway in Arctic fishing grounds near Norway to map illegal dumping by the Soviet Union. Russian authorities acknowledged that the area was a dumping ground for radioactive wastes for three decades, and the radioactive waste may include 18 nuclear reactors and several nuclear submarines. As recently as October 1993, Russia dumped hundreds of tons of nuclear waste into the Sea of Japan.

Russia secretly pumped "billions of gallons" of atomic waste directly into the earth. They said the practice of injecting the waste, which violates the accepted global standards for waste disposal and is contrary to what they previously claimed they were doing, continues today. They report that about half of all nuclear waste they ever generated has been pumped into the ground at three sites near several major rivers. They contended the practice was safe because the waste is pumped under layers of clay and shale to cut them off from the Earth's surface, nonetheless, the waste at one site have already leaked beyond the expected range.

Andrew D. Anderson

Chapter 18: Yucca Mountain in Nevada

Scientists had discovered areas of "perched water" above the water table. In addition, some Western states felt they had targeted it for hazardous facilities. The state of Nevada was expected to file a notice of disapproval if the site was found to be acceptable.

The EPA's standard was based on a new approach of using numerical probabilities to establish requirements for containing radioactivity within the repository. Cumulative releases of radioactivity from a repository must have a likelihood of less than one chance in 10 of exceeding limits established in the standard and likelihood of less than one chance in 1,000 of exceeding 10 times the limits for a period of 10,000 years.

Yucca Mountain was chosen by Congress in 1987 as a potential resting place for the nation's spent fuel rods and other high-level waste. DOE had invested $4 billion dollars testing and tunneling Yucca amid controversy as thick as the compacted volcanic ash that comprises the 1,500-food-high ridge.

The Environmental Protection Agency had ruled that DOE must demonstrate that Yucca Mountain could meet EPA standards for public and environmental health for 10,000 years. Does that mean radioactivity will not be a threat after 10,000 years? The peak radiation dose to the environment will occur after 400,000 years according to DOE. Despite objections from many scientists, the EPA decided on 10,000 years because of "tremendous uncertainties" beyond that period. Years to subside for plutonium 239, is 240,000 years.

They were looking for a stable geologic formation. One that was deep below the earth's surface and isolated from any ground water in a place that is not prone to earthquakes or volcanoes. They wanted to dig the burial chambers out of rock that is not prone to fracturing under stress such as salt, and has the ability to bind chemically with and thus immobilize any nuclides that some how manage to escape from their packaging.

The place under Yucca Mountain where a deep burial chamber would go seemed a likely site for several reasons. It was well above the current water table in a bed of tuff, an appropriate sort of rock made from compacted volcanic ash, and it had not been violently shaken by earthquakes in at least 10,000 and possibly 100,000 years, according to a study financed by the Department of Energy. It was also on the border of Nevada's Test Site, parts were heavily contaminated from years of nuclear weapons testing and must remain sealed from public use. It was not far from Las Vegas.

They were guessing what geologic changes might occur over tens of thousands of years. At the moment, the water table was low and rainfall minimal at the site, but might this change? Is it really safe to rule out the likely hood of earthquakes or volcanoes? There may be very deep deposits of gas or oil beneath the site that could tempt future generations to drill or mine into the waste burial site.

In 2002, a mild earthquake did rumble beneath the desert near the mountain chosen for a nuclear waste repository. It had a magnitude of 4.4, 75 miles northwest of Las Vegas and 3 miles beneath the surface scientists at the United States Geological Survey in Golden, Colorado said.

Nevadans Actions

In his election, the President carried Nevada in 2000, after pledging that he would oppose designation of Yucca Mountain as a temporary or permanent repository for nuclear waste unless it has been deemed scientifically safe. Nevada officials, environmentalists and their allies on Capitol Hill contended there was overwhelming scientific evidence that the government cannot safely store radioactive waste beneath Yucca Mountain without ground water being contaminated by long-term leaching. Critics also said the Energy Department had virtually ignored the risks of transporting high level nuclear waste through 43 states, within one mile of 50 million Americans providing another target for terrorists.

Las Vegas, Nevada stepped up its campaign against burying nuclear waste in the state, as the governor vetoed a presidential endorsement and activists prepared a lobbying campaign to reinforce his action. The lobbying effort was being directed by two former White House chiefs of staff, Democrat John Poddesta, who worked for President Clinton, and Republican Kenneth Duberstein, who worked for President Reagan. The campaign was to include television ads targeting lawmakers in races that could swing votes from environmentalists. Spent nuclear fuel had accumulated for decades at power plants and defense facilities in 34 states as lawmakers debated whether and where to establish a national repository.

The governor, Kenny Guano a Republican said, "Let me make one thing clear, crystal clear in fact, Yucca Mountain is not inevitable," before heading to Washington to lobby on behalf of the state's position.

The governor asked Nevada residents each to donate $1 or more to the lobbying campaign. "We will expose the Department of Energy's dirty little secrets about Yucca Mountain," he said, saying Americans have not been told of the danger of transporting nuclear waste through their neighborhoods.

Nevada's campaign would focus on lingering questions about the safety of the Yucca Mountain site and fears that the thousands of truck and train trips it would take to haul the waste across the country would lead to accidents and potential radioactive spills.

One thing was assured, no matter where a repository was placed, the large amounts of storage all across the country has to be moved by train or truck so the only arguments were what routes are used going to what part of the country.

Andrew D. Anderson

"You could have in the foreseeable future convoys or trainloads of dirty nuclear bombs rolling through your communities, past your businesses or bedrooms, and past your school for years, and that is not an exaggeration," said Nathan Naylor, Reid's spokesman.

In 1989, the Nevada Legislature passed a bill declaring the repository project unlawful. Nevada filed a legal suit against the federal government in an effort to prevent construction of the repository. Many Nevadans felt that its remote location made it ideal for use if its geologic formation was finely proved for nuclear storage. Many also pointed out that the construction and operation of the repository would bring increased employment to the state.

People living West of Yucca Mountain, nearer to the site and more likely to profit from business it might bring to the area, would support it. People living to the East, and to the East of the Nevada Test Site, were more suspicious of the project. Their homes lie downwind from the old nuclear weapons testing grounds, where higher concentrations of radioactive fallout fell during the atmospheric testing of the 1950s and 1960s. Colorado, Nebraska, and states along the likely transport routes to Yucca Mountain, also expressed concern about the site and the Western Shoshone Indians claimed Yucca Mountain as sacred ground.

The Amendments Act contained a provision that would enrich Nevada with federal monies. The state was to receive $10 million a year while Yucca Mountain was being studied and the repository constructed. Then throughout the working lifetime of the installation, Nevada was to receive annual federal payments of $20 million. If the repository served for twenty-four to twenty-eight years before being filled to capacity, the total federal payments could amount to well over $600 million.

In February, President Bush picked Yucca Mountain as the place to entomb up to 70,000 ton of spent nuclear fuel that is suppose to remain radioactive for 10,000 years. The site is 90 miles northwest of Las Vegas. A veto of Bush's endorsement was signed on Friday by Governor Kenny Guinn, a Republican, and was delivered to the state House and Senate.

Guinn's veto was allowed under rules Congress wrote for developing a national nuclear waste dump. Congress would have the final say, and a vote on whether to override Guinn was expected before August. Opponents of the Yucca Mountain plan were organizing a coast-to-coast lobbying campaign.

The Nevada Legislature passed a bill declaring the repository project unlawful. Nevada filed a legal suit against the federal government in an effort to prevent construction of the repository.

What Is This Yucca Mountain

Along its six mile length, Yucca Mountain rises 1,000 to 1,500 feet above the surrounding desert land. From that spot, you can see off to one side Jackass Flats, where the federal government has exploded some 700 nuclear test bombs over the past thirty-eight years. On one side lies the Nellis Air force Base Bombing and Gunnery Range, and a training area for aerial bombardment. One hundred and ten miles south and east of that position is the city of Las Vegas. Thirty-eight miles to the West, the Nevada and California borders meet just over the California line called Death Valley.

The Land stretching in all directions far below is wild and desolate with brush and tufts of tan grass and earthquake faults run through the area. The cones of several long-dormant volcanoes can be seen in the distance. Yucca Mountain itself was formed by violent volcanic action millions of years ago. The ground is made up of basalt, a darkish igneous rock. Below the surface and thrusting far downward are layers of tuff and rock formations that started as smoking volcanic ash.

The plans are for tunnels to be cut at a depth of 1,000 feet and to a total of more than 112 miles in length. Spreading over 1,500 acres, they would be reached by two ramps. The first would go downward at a 6-degree angle and be used to transport the waste to their final burial place. The second at the south entrance will serve as the roadway along which the rock excavated from the tunnels would be brought to the surface through a tunnel.

There will also be four shafts connecting the surface and the repository. They will act as air in takes for workers underground and will carry equipment and workers to and from the surface. On the surface, spreading over 150 to 400 acres, will be a waste-handling plant for receiving and preparing the high-level-wastes for final burial, loading and unloading docks, warehouses, administration buildings, sewage treatment plants, and a security office.

When completed, the repository will be able to hold over 63,500 tons of high-level waste, all of it contained in sealed canisters. Most of the waste, approximately 82 percent of the total, will be in solid form and will consist mainly of spent fuel assemblies from the nation's reactors. About 17 percent will be in liquid form and will come from the reprocessing of spent fuels for defense purposes. Some LLW may also be assigned to the installation.

A Look Inside The Mountain

A wide variety of studies were being conducted by 400 scientists and technicians working in four laboratories sprinkled throughout the area. Here are descriptions of some of the principal work.

The host-rock is the rock into which the repository tunnels will be carved. It will serve as the host for the waste canisters. The study of this rock is carried out by examining surface features and drilling exploratory shafts into the below-surface strata. Much of the host-rock testing was being conducted at Rainier Mesa, a region adjacent to Yucca Mountain. The work was done in a laboratory 1,400 feet underground and located at the end of a mile-long tunnel.

Visiting the laboratory, you enter the mountain and travel aboard a roofless metal train that clangs and screeches along tracks running between timbered walls. The darkness is broken by incandescent lights. There are closed passages running off to either side and signs warn that those passages are stained with radioactive contamination. A quarter of a century ago, they were used for underground tests of nuclear weapons.

The tunnel widens into a large room. This is where workers drill out samples of the rock and put them through various stress and pressure tests. The object is to determine if the host-rock is stable and strong enough to remain in place to resist shifts in the earth over the next 10,000 years and handle a complex of other problems.

Is it tough enough to keep from melting and destroying the tunnels when enduring the tremendous heat emitted by the waste? Is it solid enough to help protect the canisters against corrosive water seepage from the surrounding strata? Is it solid enough to help protect the canisters against corrosive water seepage from the surrounding strata? Water is a double villain here, not only can it corrode the canisters and cause them to leak radioactive materials, but it can also carry those materials into the surrounding strata. Then there is the question of whether the rock is solid enough to slow or prevent the advance of radiation into the strata or the outer environment. It had been said that there would be no water in the tunnels.

Rocks Tested With Intense Heat

One experiment centers on the intense heat that would come from the wastes. To see how well the rock can endure this heat without melting into a molten mass that will destroy the canisters, workers first cut an eight-foot-square block out of a laboratory wall and then drilled holes into the top and sides. Next they placed heaters, representing the hot waste, into the holes, after which pressure is exerted on the block by steel devices. Various instruments measure the temperature and pressure to which the rock is being subjected and gauged how successfully it is standing up to both. However, heat applied for six or eight hours is quite different from high nearly constant heat being applied for years.

In another test, pressure was applied to the block from various angles. Simultaneously, the block is heated to a temperature of 212 degrees Fahrenheit to drive out what water it contains. Once deprived of its water content, the block is pumped full of new water. The purpose here is to test the strength of the host-rock by exposing it to a wide variety of tortures.

Researchers have excavated more than 200 pits and trenches, drilled more than 450 holes, constructed 6.8 miles of tunnel, collected more than 75,000 feet of core samples and 28,000 other geologic and water samples, tested more than 13,000 metals for corrosion resistance, and heated 7 million cubic feet of rock. This information was gathered for building a simulated mountain model.

The rocks are made up of tuff, a type of rock that began as burning ash when the area was hit by violent volcanic action between 13 and 18 million years ago. It is a rock that can be porous or nonporous or a combination of the two. Beneath Yucca Mountain, it divides itself into four major layers that extend to a depth of 6,500 feet. The DOE plans to situate the repository in a layer made up chiefly of the nonporous type, which is called densely welded tuff.

Following that long ago volcanic action, the densely welded tuff cooled and solidified more slowly than the porous type, known as non-welded tuff. Consequently, it is less crumbly and far stronger than the non-welded variety. It is so strong that it has a compressive strength twelve times that of concrete and will give construction crews trouble when they try to drill through it. Though it contains some porous areas and though it is marked with many cracks and crevices, it promises to do well against heat, the invasion of corrosive water, and shifts in the earth for the next 10,000 years.

It is especially vital that the rock be solid enough to lock in any radiation leaking from a corroded canister. The canisters are expected to resist the effects of corrosion for a period of up to 1,000 years. After that length of time, leakage is to be expected. The rock will have the job of keeping the radiation in place and "slowing" or preventing its escape into the environment.

The densely welded tuff is expected to do particularly well in one respect. It is known as a good conductor of heat. This will enable the rock to absorb and pass on the heat from the waste without melting and becoming a dangerous and destructive molten mass. In addition, suppose underground water seeps into its cracks and crevices, its fractured structure will allow steam to escape toward the surface and dissipate if the water, heated by the waste, ever reaches the boiling point. The waste canisters would be protected from the steam's corrosive effect, but for how long?

Until recently, scientific theory held that salt formations would serve best for a repository, because it was thought that their ability to seal fissures in their structure would efficiently lock in any leaking radiation. The federal repository for reprocessed HLW in New Mexico was carved into a salt formation and the leakage problems are suffering and casting serious doubt on this theory. The welded tuff at Yucca Mountain is now widely considered superior to salt as a host for the wastes, but what does, "widely considered superior" mean and will it be tested for its durability?

Water Studies

Death Valley, the driest area in North America is lying just thirty-eight miles away, but the Yucca Mountain region is not hot and arid. It receives and average of just three to six inches of rain annually. Yet, despite this scant rainfall, the DOE studies focused much of their attention on the dangers of water to the repository.

Though the densely welded tuff promises to help protect the tunnels against water seeping in from the surrounding earth, these studies are considered of vital importance because of the possibility of flash floods and future changes in the area's climate. Should something happen that allows water to come through the host-rock and corrode the waste canisters, it would release a radiation that could then work its way up to the surface or down to an aquifer that lies 2,000 feet below the mountain. An aquifer is an underground source of water embedded in porous rock. The aquifer would likewise be threatened if liquid wastes escaped from their canisters and oozed down toward it.

It is to help reduce the dangers of radiation leakage that the repository is planned for at a depth of 1,000 feet. A depth that is exactly halfway between the surface and the aquifer. The area at this level is called an unsaturated zone itself and is without any appreciable amount of water. It is below the reach of the water near the surface, and above the reach of the water in the aquifer. Yucca Mountain boasts and unsaturated zone of three Western sites.

The average three to six inches a year and because much of its moisture is evaporated by the area's heat, only a scant amount of water will ever manages to penetrate the earth. The scientists estimate that, of that small amount, only 0.02 inches of the annual rainfall will ever work its way down to the 1,000 foot-deep repository. It will not be enough to do any harm, but suppose one waste canister does corrode or break open? The scientists claim that the aquifer will be well protected. Should either solid or liquid waste escape, each would take about 10,000 years to inch down to the level of the aquifer a study said.

Chapter 19: Simulation of Storing Nuclear Waste

How Will It Be Packaged?

Eighty two percent of the wastes will be in solid form with most being spent fuel assemblies. Fuel assemblies are made up of the stainless steel or zirconium rods into which uranium pellets were inserted for fission. They measure from three to fourteen feet in length and are about a half-inch in diameter strapped together in bundles of 30 too 300 rods. They will be packaged in casks at the power plants. Casks are containers the wastes will travel in.

The wall of a typical cask will be six to eight inches thick and divided into three parts. There is an outer shell of stainless steel and an inner shell of stainless steel. Between the two is a layer of shielding lead to lock in the cargo's radioactivity. All casks must be designed so that they can survive the most severe road, rail, and water accidents without splitting or breaking open releasing hazardous radioactivity.

In testing, a cask was dropped from a helicopter. It hit the desert floor at 235 mph and dug itself four feet into the hard-packed soil. The only damage suffered was some scratched paint. For punctures, a cask was dropped more than "three feet" onto the up thrust end of a steel bar six inches in diameter and 8 inches long. The cask was dented, but suffered no splits or punctures.

For a fiery crash, a cask was strapped aboard a railroad car and sent crashing into a thick concrete wall at 81 mph. Both car and cask were then engulfed in a jet fuel fire that lasted 125 minutes. After about 100 minutes of intense burning, a small crack about as thick as the edge of a dollar bill appeared in the cask's outer skin and some molten lead from the shield oozed out. The fire caused the lead shield between the cask's inner and outer walls to melt. The simulated radioactive materials deep inside the cask remained where they were.

In Great Britain in 1984, a cask was placed on the flatbed of a truck and rammed by a diesel locomotive. The locomotive was pulling three 35-ton cars and moving at 100 miles an hour at the moment of impact. The train was destroyed, the cask hurled 200 feet along the tracks, but sustained only minor scratches.

Liquid radioactive waste from the government's nuclear weapons program will be mixed with sand and converted to glass before being sent to the waste repository. The standard shipping receptacle will be a stainless steel cylinder 10 feet long and 24 inches in diameter.

Method of Transporting Waste

Responsibilities for the transportation of the systems will be the Department Of Energies, Department of Engineering, Office of Civilian Radioactive Waste Management, The United States Department of Transportation, and the Nuclear Regulatory Commission.

Don Hancock, the nuclear waste program director for the Southwest Research and Information Center in Albuquerque, noted that a freight train carrying hazard wastes last year in a tunnel in the state of Baltimore, Maryland caused a fire that burned for five days. They had to close the tunnel. He said the propane fire burns at 2,000 degrees F.

The Nuclear Regulatory commission specified that casks be tested by burning them in fuel for a half hour at a temperature of 1,475 degrees F. That standard was adopted in 1965.

For water-tightness, a cask was submerged in three feet of water and left there for 8 hours. It was then lowered to a depth of 50 feet and left there for another 8 hours. When it survived the accident, fire, and water tests in sequence it was certified for use by the NRC and its mass production will begin. The certification lasts for five years with inspections along the way to make sure the containers are performing as expected.

The wastes will be moved to a Monitored Retrievable Storage, MRS, at the repository by truck or rail where the workers will remove the wastes from the casks and repackage them in canisters used for burial.

Trucks will carry the units that will weigh between 25 and 40 tons and will be built to hold from one to seven spent fuel assemblies. Larger models weighting up to 200 tons are intended for the railroad cars that will travel to the repository and can accommodate thirty-six or more assemblies. They will be tubular in shape, and can range up to 12 feet in diameter and 22 feet in length. The wastes will be sealed in casks and placed aboard a train that will be used only for the transport of nuclear materials. Railroad cars carrying HLW are not to be attached to trains hauling general freight. Trucks will carry waste accompanied by military vehicles. The travel by rail will have buffer cars fore and aft of the waste to shield it.

When the wastes arrive at the repository it will go to the MRS which will be a surface - level receiving-and-handling station to prepare the wastes for burial. This job requires its own personnel, a special building, and special equipment, and is more economical and convenient to have the manpower and MRS located at the repository. The building will be able to keep the wastes in storage until the repository is ready to handle them.

Final Burial

At Yucca Mountain present DOE plans call for the waste to be fused with a protective material that will help to lock in their radioactive contents more securely. The material to be used is borosilicate glass, an extremely strong substance that can easily be produced on a large scale. The borosilicate glass will be in molten form when the wastes are mixed with it. Workers will do the mixing by remote control to safeguard against radioactive contamination.

While still in a molten state, the glass and it wastes will be poured into stainless still canisters, called pour canisters. When the mixture has cooled and solidified, the poured canisters will be plugged and welded shut. They will be tested for leaks and decontaminated. Decontamination will remove any radioactive residue from their outer sides.

The poured canisters will then be inserted into burial containers. These containers are to serve as partitions between the waste-filled pour canisters and the repository environment. They are to be constructed of carbon steel, stainless steel, or copper based alloys. Each of these metals resist corrosion if exposed to such geologic condition as the invasion of water into the repository tunnels. Only when the burial containers have been closed and sealed will the waste be moved into tunnels for permanent storage.

When the capacity of some 63,500 tones of waste is completed, the tunnels will be filled with dirt, sealed off, and then constantly monitored for any sign of escaping radiation. The surface area at the Mountain will be decontaminated and allowed to return to its original state.

Conclusion

The first years of the 21st century we seldom saw the sun shining all day, the weather has more tornados and floods, and our production of food needs to increase. There are a large number of people all over the world that do not get enough water to drink and die, and our water all over the world is decreasing for irrigation and drinking, the land we irrigate all over the world is becoming saline, and the air we breathe is becoming more deadly. What kind of future will the world have?

Since we live in a busy aggressive world with a slow economy, the intent of this book is to present shared and new background information and provide a perspective on some of the pollutions we endure twenty four hours a day. It is meant to assist anyone to get a better working knowledge of this polluted world as it grows more polluted at an unstoppable rate. The idea is to present this information to a wide range of readers by discussing issues in a variety of prospectives that show the struggle between humans, pollution, and human predators such as microbes that prey on our human body.

Andrew D. Anderson

Select Bibliography

Books

Andryszewski, Tricia. "What To Do About Nuclear Waste." *The Millbrook Press.* Brookfield, CT.

Campbell, William F. & Hynes, Thomas J. "The Agriculture Crisis," A critical analysis of the agricultural policy of the United States. *National Textbook Company.* Lincolnwood, Illinois.

Doland, Edward F. & Scariano, Margaret M. "Nuclear Waste The 10,000-Year Challenge." *Library of Congress,* Cataloging-in-Publication Data.

Flaningam, Carl. "The Future of American Agriculture." *National Textbook Company.* Lincolnwood, Illinois USA. 1975.

Jacob's, Nancy R. & Landes, Alison. "Garbage and Other Pollution." Wily, Texas.

Jacob's, Nancy R. & Quiram, Jacquelyn F. "The Environment, A Revolution In Attitudes." Wily, Texas, 1998.

McCuen, Gary E. "Protecting Water Quality." *Gary E. McCuen Publications Inc.,* 1986, Hudson, Wisconsin 54016.

Miller, Willard, & Miller, Ruby. "Contemporary World Issues, (Environmental Hazards: Air Pollution)." *ABC-CLIO Inc.* Santa Barbara, CA.

Magazines

Ackerman, Jennifer. "Food: How Safe? How Altered." *National Geographic,* (May 2002). "A Lot of Beef." p1, "How Altered." p33.

Alvarez, Robert & Makhijani, Arjun. "Radioactive WASTE." *Technology Review,* (Aug/Sept. 1988) p42.

Barnett, Megan. "Investigative Report: Making A Stink," *U.S. News & World Report,* (August 5, 2002).

Cowley, Geoffrey. "Are Organic Foods Really Better for You?" *Newsweek,* (Sept. 30 2002).

Holloway, Marguerite. "Blue Revolution." Farm Fishing, *Discover,* (Sept. 2002).

Klesius, Michael. "State of The Planet."

Long, Michael E. "Half Live." The Lethal Legacy of America's Nuclear Waste. *National Geographic,* (July 2002).

Mitchell, John G. "Down The Drain."

Montaigne, Fen. "Water Pressure." *National Geographic,* (Sept. 2002).

Montaigne, Fen. "Atlantic Salmon." *National Geographic,* (July 2003).

Reisner, Marc. "The Next Water War: Cities Versus Agriculture." *Issues,* (Winter 1988-99).

Sheelwright, Jeff. "Yucca Mountain: A Safe Nuclear Wasteland." *Discover,* (Sept. 2002).

Shute, Nancy. "The Weather Turns Wild." Global Warming Is A Misnomer. *U. S, News & World Report.* (Feb. 5, 2001).

Spake, Amanda. "Don't Breath The Air." *U.S. News & World Report.* (July 1, 2002).

Newspapers

Arnold, David. "Could global warming produce a big chill?" *The Boston Globe.*

Arnold, David. "Ozone making air unhealthy in New England." *The Boston Globe.*

Bagrov, Yuri. "Glacier breaks apart, buries Russian village." *The Boston Globe.*

Crenson, Matt. "As Ocean Swallows Land, LA. Prepares." *The Boston Globe,* 22 August 2002.

Healy, Patrick. "Warming waters." *The Boston Globe*, 30 August 2002.

Holland, Jesse J. "Company recalls 19m pounds of beef." *The Boston Globe,* 20 July 2002.

Lawless, Jill. "Asia smog affecting climates study says." *The Boston Globe.*

Nessman, Ravi. "A vow to save the Dead Sea: *The Boston Globe,* 2 Sept 2002.

Ready, Tinker. "The war against the sea." *The Boston Globe,* 5 June 2001.

Smith, S. "Pneumonia in Bay State resists drugs at high rate:" *The Boston Globe,* 3 Oct. 2002.

Ward, Diane Raines. "Our water troubles are just beginning." *The Boston Globe.*

Andrew D. Anderson

About the Author

A retired Aerospace Engineer, he is concerned that the sun is seldom seen hidden by clouds that carry microbes, bacteria, human feces, and toxic chemicals small enough we cannot exhale. He is concern about an estimated 1.2 billion people that drink unclean water and 2.5 billion who lack proper sewage systems. Millions of ton of raw sewage, chemical waste, fertilizer, and animal feces kill our fish and pollute our drinking and ground water. Thousands of tons of nuclear waste leaked into the ground and atmosphere.

He studied the growing shortage of water across the world from over pumping our acquirers and farmers waste water irrigating their land using inefficient sprinklers and digging and using new wells.

www.ingramcontent.com/pod-product-compliance
Lightning Source LLC
Chambersburg PA
CBHW031124180526
45160CB00001B/13

* 9 7 8 1 4 1 8 4 5 4 1 8 0 *